Also by Mark Kramer
Mother Walter and the Pig Tragedy

THREE
FARMS

THREE
FARMS

Making Milk, Meat and Money from the American Soil

BY MARK KRAMER

An Atlantic Monthly Press Book
Little, Brown and Company   *Boston/Toronto*

FIRST EDITION

Portions of this book, in slightly altered
form, appeared in *The Atlantic* and *Country Journal.*

Library of Congress Cataloging in Publication Data

Kramer, Mark, 1944–
   Three farms.

   "An Atlantic Monthly Press book."
   1. Agriculture—United States.   2.   Farm life—
United States.   3.   Farmers—United States.
I. Title.
S441.K72   338.1'0973   79-25920
ISBN 0-316-50315-0

ATLANTIC–LITTLE, BROWN BOOKS
ARE PUBLISHED BY
LITTLE, BROWN AND COMPANY
IN ASSOCIATION WITH
THE ATLANTIC MONTHLY PRESS

BP

Designed by Janis Capone

Published simultaneously in Canada
by Little, Brown & Company (Canada) Limited

PRINTED IN THE UNITED STATES OF AMERICA

for Ray Totman

Introduction

In suburbia, where I grew up, I always ate my fill. I felt kindly toward the farmers who must have grown my food, but I thought little about them. Twelve years ago, in my mid-twenties, I moved to the country, to a run-down, north-slope, sodden, clay-soiled Massachusetts farm too small and too infertile for any farmer today — even the most old-fashioned sort of Yankee side-hill farmer. I began to pay attention to farmers' business of raising and selling food, and I was puzzled.

Here were well-meaning people laboring hard on good land, whose animals looked sleek, whose crops grew lushly, whose thinking about their work seemed rational and fascinatingly knowledgeable. And yet, one after the other, my neighbors sold out and took factory jobs. This book has grown out of my initial curiosity about why these family farmers can no longer survive. At its most general level my curiosity was about the social and political effects of changing technology. It turns out that who raises our food, who invents, finances, makes, and distributes the hardware and materials that go into food raising, who owns the farmland, who ships, who processes, and who sells the food that farmers raise are all questions of the utmost social importance.

A siege may represent the most extreme form of political control that can be achieved by affecting food supply, but it is not the only one. Changes currently taking place in America's food system are shifting into fewer hands the power to decide who pays how much for what food. To understand the source of one's next meal is to understand one's own political vulnerability.

This book is about people, politics, and food. It concentrates first of all upon the people growing the food we eat and, through them, upon the technology and business structures that shape their lives and ours. For the first time in American history the possibility exists of less than free market control of our food supply. New technology and new business structures provide opportunity for increased control by some companies and institutions.

We, as citizens, are just beginning to experiment with the possibility of refusing some technology and some business structures — and of choosing instead what seems more to our advantage. We have grown critical of nuclear power plants, supersonic passenger planes, excessive dependence on cars, what TV does to kids. Similar criticism of the new agriculture is as vital to our common survival. But it is hard to understand the complex system that keeps us fed. Our meals, these days, reach us after traveling a long road of supply that begins with farmers, fuel, seeds, and tools and ends with room-sized ovens, supermarket-chain purchasing agents, quarter-mile-long deep-freezes and checkout persons. Ultimately, the best way I know to discover the politics of our food supply is to remember the people involved, to see them within the system, to see the whole thing operating. That's what I have tried to do here, starting and ending with the people who grow our food.

The subject is a writer's dream and a sore temptation as well. A pervasive wistfulness comes ready-made with the topic of declining family farms. Farming has shifted polarity, gone from cozy to harsh within easy memory of many still in the trade — first-rate poignancy. In the course of my writing I found myself tempted to run with the prevailing wind of sentiment, to blame, to use the force of this public loss to help ridicule some of the public devils of the moment. In the end I have tried to leave complex situations complex. One of the more confusing aspects of our current agricultural situation is that many who stand to lose the most by the current run of technical change are the heartiest advocates of such change.

My choice of particular subjects reflects the refusal of the real world to bear out easy theorizing. The most poignant of all situations in our agriculture, from my regional perspective, is the recent devastation of New England's farm population, but I examine it from the perspective of a prospering family of survivors. The Iowa corn-hog farmer I portray is dedicated to the concept of family farming and is a champion of collective action by farmers, yet he is driven by his exposed position in a fast-changing business to consider new strategies that contradict his values. The California corporate farm I write about is an awkwardly, perhaps a harmfully large operation, but its people are full of admirable determination in the face of a series of business reverses. In fact, what the Yankee, the Midwesterner, and the transcontinental businessmen have in common is their ability to apprehend a system that nowadays makes victims of its slacker participants, and to operate with the canniness and vigor needed to make do in hard times.

I have approached the lengthy task of assembling a book about these people with no more pretense of objectivity than the heartfelt desire to be accurate. Just to spell out my own bias, my utopia offers opportunities for most people to do work that makes them pleased with themselves. If this preference doesn't show through the rural haze, I haven't said my piece.

In the course of four years of work I have received guidance and support from many people. I am sure I am neglecting some whom I ought to include, for which I beg pardon in advance. The manuscript's errors are mine; its finer moments grow from the generosity of others. I am most of all indebted to the late Raymond Totman, a farmer's farmer and a poet's poet, whose chiding but affectionate counsel improved my aim throughout the period of researching and writing. How I wish he might have seen the book in print!

I also wish to thank the Rockefeller Foundation, Dr. D. Lydia Brontë in particular, for the kind award of a Humanities Fellowship. Dr. John Foster, Chairman, Department of Food and Resource Economics, University of Massachusetts, deserves much appreciation for taking in a stray journalist, offering students, advice, a letterhead, and the stamp of officialdom to a project bred in wilder country. Tracy Kidder lent order and courage when chaos loomed. Deb Robson turned the resulting cryptographer's puzzle into clean, copyedited typescript with skill and care rare in today's workplace. Finally, I acknowledge the strongest of personal debts to Sarah Ellsworth and to Dick Todd, my editor, in both instances for graciousness, hospitality, insight, and wit shared freely. Others as deserving of thanks include the persons described in this book and Jim Autrey, Berge Babullian, George Ballis, Harper Barnes, Tony Bell, Roger Blobaum, Tony Borton, Vance Bourjaily, Robert Buck, John Callison, Iris Cavagnaro, Harvey Chandler, William Chester, Eliza Childs, Robert Christianson, Jim Cromack, Jesse de la Cruz, Hugh Davis, Elton Epley, Robert Erburu, Peter and Maralyn Fuchs, Susan Goldhor, Larry Gross, David Hartwell, Warren Henniger, Kris Herzig, Stanley and Pauline Herzig, J. B. Jackson, Lindsay Jones, Dorothy Kalins, Fran Kidder, Michael Klare, Sally Kleinfeldt, William Knapp, Sidney and Esther Kramer, Al Krebs, John Lancaster, Jon Lagreze, Robert Light, Mary Lippincott, William Lockeritz, Sidney Lyford, Judith and Terry Maloney, Jim Manwell, Dick and Elsie Mayer, Mary Mazzei, Kristan McInnis, John Paul McNutt, Paul and Mary McNutt, Conn Nugent, Sue Parilla,

Katy Peake, John Perkins, Wendy Posner, Erliss Rohret, Barbara Rosenkrantz, Idella and Glen Schmidt, Bill Schreiber, Daniel Schwartz, Lee Searle, Denise Seldin, Morris Sherman, Marty Strange, Susan Todd, Miriam and Peter Weinstein-feinstein, Ross Whaley, Edie Wilson, Fred Winthrop, Rick Wolff, George and Susan Wolk, Mark Zanger.

Making
Milk

I

L ate in the summer of 1910, shortly after his ninetieth
birthday, an ailing old Yankee, Joshua Totman by name,
feeling caged by the narrowness of his round between sickbed
and dining room in his son's farmhouse, peered out the
window at the high, even and lush greenness of a field of
second cutting hay just across the road. He proclaimed to his
son, "Frederick, I want to mow again. I want to mow once
more while I still can."

According to Totman family lore, whether Joshua still
could was already an open question. Nevertheless, Frederick
went about preparing for the chore. The mowing machine
was especially dear to Joshua. Until he was in his sixtieth year,
like his forebears for generations past, old Joshua had harvested
his hay by hand, with scythe and hayfork, together with the
gang of neighbors needed to get the job done. Starting with
a different farm every year, they moved on to return the
favor at every farm in the neighborhood, laboring hard by
day and eating together in the evening.

In the 1860s, country newspapers were filled with news of
trials of a progression of mowing machines. The reports grew
more and more positive. There were problems — clogging,

protecting blades from obstacles — but these were getting solved. Joshua was said to have bought the first machine in Conway, Massachusetts, perhaps not until the late 1870s. The purchase freed him in a way — placed him outside the constraints of neighborly interdependence. His farm prospered and the news spread. The neighbors could not have afforded to wait too many years before they followed suit and equipped themselves to mow on their own. Joshua, however, had been first, and Joshua loved mowing.

The old man was ailing but he tottered out of doors to supervise as Frederick (who was fifty-four) hauled out the mowing machine and harnessed up the team. As soon as the horses were in place, Frederick himself climbed up onto the cast-iron seat. He drove the horses across the road and, after pausing for a second to observe the rolling field, moved in a way through the high grass, then engaged the cutter bar and mowed around a long and perfect rectangle, defining for his father the easiest mowing on the farm — two acres, flat, free of trees and ledgy outcroppings, squared to eliminate the point rows of irregular fields. Coming back to the dooryard, Frederick stopped to chat for a moment with a farmhand known in family lore as "old Deck," a quiet and usually temperate man somewhat younger than Joshua, who, during Joshua's decline, had been doubling as his nursemaid. Frederick instructed Deck to walk behind as the old man mowed, "so close that if he topples off the seat, you'll get to him quick enough to set him up straight again." These precautions taken, Frederick climbed down off the mower, helped Joshua up in his place, and watched as the procession headed back for the mowing — Frederick's team of white horses in the lead, then Joshua, seated now with reins in hand, then Deck striding behind, and, to round things out, one nine-year-old boy who was excited to see his grandfather in action. The boy, an old man now, says he would be surprised if a cowdog named Topsy hadn't been trailing after him at the tail end of the parade.

Frederick's compliance to Joshua's desire to mow represented an act of deference, of filial duty in the face of ingratitude, of the setting aside of an argument that had, eighteen years earlier — when the old man was already seventy-two — stopped father and son from farming together.

When Joshua had moved to Conway, many years before, it was not to the farm where the mowing scene occurred, but to another farm in a section of town called Shirkshire, about five miles farther north, on the banks of the tiny Bear River. Frederick was one of six surviving children who were raised on that farm. Of the six, he was the only one who elected to stay home and farm with his father. Frederick must have been a patient young man, because it was not until he was in his late thirties that he ran into trouble with Joshua. Frederick brought up the subject of his future one day, and father and son parted ways on the question of inheritance. Joshua insisted that upon his death the farm be shared equally by all his children. Frederick insisted that since he was the son to stay home and farm, adding, after all, materially to the worth of land, stock and buildings, he should be in line for half the farm, and that his brothers and sisters should take shares from him for the value of the rest. He wanted to own the ground he farmed.

The entire Shirkshire quarter section might have been worth four or five thousand dollars, but however cheap prices were then, cash in the last century was a scarce resource to hill farmers. Frederick bought a farm a few miles down the road just above the village of Conway on the Bardwell's Ferry Road. It was heavily mortgaged, and even so, Frederick could only afford it because it was a farm that had been allowed to run down. The barn had burned too. The first thing Frederick did upon taking title was to buy an old barn in Greenfield, fifteen miles away, tear it down, ship it on the local railway to Conway, and draw it piecemeal up the steep hill from town to the farm with a team of mules.

Frederick's new farm was a fortunate purchase. Although

family history doesn't estimate how much the enticement of the beautiful site at the fork of the South and the Deerfield rivers added to Frederick's sense of independence, it must have influenced him. For the new farm was — and remains to this day — one of the most favored in the entire county.

There was a barn raising. The neighbors came and in one day resurrected the Greenfield barn in Conway, then ate a big and festive supper together and went their separate ways — nearly all had mowing machines by 1892.

Old Joshua stayed away from his son's new farm, minding his own business up on the Shirkshire quarter section until infirmity finally forced him to move in with his son. He spent his final year on that farm, and was out mowing only months before his death.

The old man didn't topple that day. He mowed, around and around the hayfield Frederick had squared up for him, doing something useful, celebrating routine — which could go on and on — on a late summer's day in his ninety-first and final year of agricultural life.

The boy who had run alongside the mowing machine, close in behind the farmhand Deck, was Frederick's baby, the youngest of his six surviving children, Joshua's grandson, Raymond Totman. Raymond was to take over the place in his own time, farm it for half a century and pass it on to his own son. It's wonderful to talk to Raymond about his grandfather Joshua, like talking to a friend about someone you haven't yet met, except that in this case the someone was born a century and a half ago, recalled the election of Andrew Jackson, and had spent boyhood days cutting away at what remained of Massachusetts's virgin forest. When Joshua died, near the end of 1910, his farm in Shirkshire was logged off, then sold out of the family to two Polish brothers named Rutkowski. Raymond's father got a fifth part of the purchase price and no more, as per Joshua's wishes. What Raymond remembers best are not stories of past elections, big trees, or the lost farm, but the efforts his father took that final season to make *his* father happy.

It is September again now. We are in the same farmhouse, in the same dining room. Raymond is lounging. He is reclining, stretched into an elongated Z in a large brown plastic-covered easy chair. He is short, seventy-five, is blessed as were his father and grandfather with a full head of hair and vital pale blue eyes. Also like father and grandfather, he is built solidly, like an inverted trapezoid, broad end up, narrow end down. He looks like a man who was rugged once.

"You'll see," he is warning me, "you get to a certain age and your chest falls. Makes you look smaller."

Raymond is alternately staring at me and out through the window into the dooryard, where his eldest son, Leland, who has stayed home to farm, is at work on the engine of a big John Deere tractor. Raymond has just finished telling me about Joshua's final mowing. He is silent for a full minute. Behind Leland's tractor and on across the road we can see the mowing where the scene occurred. This is corn chopping season — time to make silage — and there is a corn-chopper hitched on behind Leland's tractor, a bulky green metal housing sprouting silvery sheet metal tusks and a yellow snout like an elephant's trunk that trumpets high in the air.

Raymond points out the window to the lush seeding of alfalfa and timothy.

"Wants mowing too. Been mowed three times already, and I expect I'll cut it again — " he is careful to add " — when Leland says the word. We mow hayfields more times each summer than old Joshua would have. Gives a much better quality of feed, you know. They were after bulk, we're after quality. Different times, different things you have to get from hay. Then it was a matter of keeping the stock alive and fed through a long winter, doing it all by hand, with horses, of course. Now it's business. You have to feed quality or there's no sense of feeding livestock at all nowadays. Then it didn't matter so much."

Another pensive moment. Then he says, with a sigh, "Thank goodness I can still mow — and I can even manage it without a farmhand running along behind me, though you might not think so to look at me. That wasn't always the case, you know.

"I had cancer back in ninteen sixty-one — throat cancer. It was treated with X rays and I have been fortunate enough to survive. Leland took over then, and I've been helping out ever since.

"I expect Lee had an easier time than my dad did with his father. That sickness may have eased the transfer of power — the winter I was sick Lee ran things, and he's kept running things since. I'd say I was ready to pass over the reins without much prodding or regret. Still, there have been times — no, I had other interests. I was assessor here for years, and I was a selectman for three years, was also on the boards of the library and the bank."

"Had you," I ask him, "the same easy time taking over from your own father?"

He answers, eventually, with some deliberation and at some length.

"I went off to Worcester Tech in nineteen twenty-one, took the 'short course.' I'd told Dad that if he and my brother Bill didn't get along I'd come back. They didn't. The professor of English at Worcester was an ex-newsman, used to sit looking bored and half-asleep when he talked. He got going one day on what we all wanted to become. 'If your aim is to make money, go away,' he told us, 'because the more knowledge you get, the more you'll want to get.' With that I came home.

"Frederick was all of forty-five when I was born. I was a late child. He died at seventy-one in the year nineteen twenty-six. So he was ready to slow down when I was just back from college. Still, now that you mention it, he did indeed take a little prodding.

"I came back from Worcester. I worked with him for a few years, and he didn't pay me — when I wanted money I had to apply to him for it. We didn't have any definite agreement. I began to wonder about that time what the dickens to do about getting married. I thought about it quite a lot, but I couldn't see any alternative, any way out at all. So I decided to take the bull by the horns.

"We were working on wood that morning. I got my courage up and I told Dad I was getting out. I gave him notice. I didn't like it one bit, but I told him I was leaving. And you know what? Nothing happened. I brought the matter up a second time a few days later, and it was then we came to an agreement — I should have some money, about a hundred dollars cash and his note for four hundred more, plus a partnership on the farm.

"Years later he told me he was greatly relieved, that he'd been worried he'd have to call in the auctioneer. He said he remembered the hard time Joshua had given him, and he didn't ask as much as he might have been entitled to from me for my share when I bought in. He offered me half the equity if I'd come in with him, and would also promise to take care of Mother and him as they got older. A few years after my father died, my mother followed him.

"I borrowed from a bank and paid off all but two sisters, and they each took personal notes for their shares — six, seven hundred dollars a share it came to — and I was in business and I was able to stay. I've been in this house all my days, owned it since five/nine/twenty-seven. When Leland took over I remembered my own father's hard experience with Joshua, and his kindness toward me. I had the same feeling regarding Leland."

The elder Totmans' dining room is a room for living; Raymond and Mildred pass a good part of each day there. The room runs the width of the house. Ray sits in his chair by the south window, looking out over the farm he built. Mildred's rocker, empty just now, faces the north window, looking out on the birds, on a sheltering shrub and berry garden whose posts and trellises are stacked with bird feeders and birdhouses, bird perches, and, in every season of the year, with birds. In the distance, past the birds, down the hill and on past Leland's house, across the valley and most of the way up the side of a small mountain five miles to the north called Patton Hill ("up past that clearing, follow the line of the road around. Now, see that white spot? That's the place"), is a schoolhouse where Mildred taught, half a century ago, just before she married Raymond and came to live on the farm.

Mildred is in the kitchen now; I hear the clattering of the cast-iron stove lids as she adds more wood to the fire. The kettle whistles, teacups rattle, and then she serves high tea, pouring with a shaking but sturdy hand, offering raspberry tarts, beautiful thin tarts of flaky lard crust and jam from the bushes beyond the bird feeders. Mildred is one of the last of the old-time Yankee farmwives, slight now, and shy in her gestures too, with graying hair and almost translucent skin. Age has not slowed her down. She worked on the farm when things were tight, years ago, driving truck and tractor, feeding the workhorses, shoveling out stables. As soon as they could afford a farmhand she stopped doing that. She says she has never been a "sitter-around," and has worked as hard indoors as she had at fieldwork. In the normal routine of the farm she hasn't been needed for outdoor chores in many years, and she construes her own farm partnership in terms — terms far too modest, one suspects — that would appall a modern feminist. "Ray and I had an agreement, and we had things divided up nicely all our lives — his business was the farm and mine was the house. I never consulted him when I bought a new refrigerator or a new range or furniture. We raised the kids together. He always handled the farm. We both tried to do a good job."

Mildred still has the Ladies Aid over on occasion to plan and perform good works in the community and to enjoy each other's company. "Most of the time, though," she explains, "we're alone together in the house now, right here in this room. We have talked for so many years we've about run out of things to say, sometimes. But — " Mildred waits out a well-timed moment " — we don't have much left to argue about either." She sits now in the rocker, chatting easily and looking out at her bird feeders, a copy of *Peterson* and a pair of field glasses on the lampstand next to her. "We've had a few new ones lately," she says after a while, "ones you don't see around here frequently. Housefinch. Maryland yellowthroat."

Ray mans his station across the room. Nestled in his recliner, gazing out the other window into the dooryard, he watches abstractedly as Leland slithers out from under the big John Deere, turns the key, leans over the instrument panel, nods with satisfaction. Leland disappears from the window frame lugging an armload of tools, reappears a moment later empty-handed, climbs on the tractor. The roar of the motor is the first sound that penetrates the quiet room from without. The sound abates as Lee drives down the tarred roadway past maples already old enough to have been tapped when Frederick bought the farm in 1893, across the Bardwell's Ferry Road and into the beautiful acres of cornland on the other side. The tractor moves down a farm road, then disappears behind tall rows of field corn. Soon through the window comes the sharp snarling sound of the corn-chopper. After a while Lee has cut into silage the tall rows of corn that have hidden him from view, and we see him again as he finishes the final rows of the upper piece, then turns downhill toward the next cornfield. Raymond nods as he sees the tractor emerge. He clears his throat, preparing to make pronouncements.

Young Yankees are by and large a quiet breed. But when they grow old, a few change into pleasingly discursive sorts. Others grow quieter yet. Raymond is the talkative sort. His speech has a touch of Victorian ornateness, a solid and economical core of meaning adorned with garlands of embellishment. He is given to using words like *contention* instead of *opinion*, and *frivolity* instead of *nonsense*. It is an act of charity toward fortunate listeners, grounded in a pre-TV attitude about the kind of "visiting" we are now engaged in, in the dining room. "Talk is cheap," Raymond explains, "and is a distraction when there is work to do. But on the other hand, talk is thrifty entertainment." Raymond does it well, as he has done most other things. We talk of talking, and he tells the story of chatting with a garrulous insurance salesman once, years ago, during sugaring season.

"When you're boiling sap, it's quiet, you have to stay in one place, and there's not much in the way of suitable distraction going on," he recalls. "I don't know just how he got onto me, but there was a life insurance salesman canvassing the area for likely prospects, and he came straight on into the sugarhouse, just when I was wanting someone around for a visit. He was a nice enough chap, and besides talking about insurance, he told quite a few good stories. We passed a pleasant enough afternoon together, talking our fool heads off. Then he says, 'How about that insurance?' and I say, 'Come on back in a month's time, when I've had a chance to consider the matter a bit.' Well, he was as good as his word and a certain period of time later he showed up again. By then I was all done sugaring. Had the buckets washed and stored, and I was busy doing spring plowing. You know when you get to that stage of the year, there isn't a farmer around is going to stand still for more than a minute at a time. That salesman showed up, and I'd be surprised if I said a dozen words to him before I took off plowing. He never showed up again but we'd had quite a good conversation back in the sugarhouse. I still remember it well." Raymond is laughing now. He loves to talk. We talk, and watch Leland cut corn.

I get in the habit of visiting often. Raymond has time to spare now, although half a year later, in spring plowing season, he is as short with me as if I were the salesman incarnate. But for the most part we entertain each other. I savor the vastness of his knowledge of growing things, of old ways, his acceptance of his lot, his courtliness, which I learn goes deep. I bring him the outside world, allow him to test out the knowledge he has induced about the world during years of industrious isolation. I confirm that he's not behind at all in his thinking or information and it pleases him again and again. He is an exacting and persistent questioner. It does not pay to advance half-baked or tentative positions in his presence. He suffers fools at some length before interjecting a delicate question. He makes one cautious.

Raymond tells about himself. He hasn't traveled much, hasn't much taste for it. From 1922, when he came back from the short course at Worcester, until 1951 he seldom strayed beyond the confines of Massachusetts save for anniversary tours to Washington, D.C., in 1943, in the middle of the war, and to Florida in 1951.

Later that same year Raymond was named number two farmer in all of Massachusetts, winning a sort of agricultural Oscar called the Green Pastures Award. Mildred has pasted in the family scrapbook a photograph of Ray at the Awards Dinner, down at the big Eastern States Exposition fairgrounds in Springfield. It's a dark snapshot, glossy, with a wide, white, ragged-edged border, in the manner of the fifties. It shows a slight but cocky-looking fellow with a stance like Mickey Rooney's and the kind of absent expression one might expect to find on a Marine standing presidential guard duty. Ray is surrounded by professors of agriculture half a head taller and half as broad as himself. He seems uncomfortable, out of his element in boxy pin-striped suit and striped silk tie.

On the facing page of the scrapbook is a picture taken that same summer of Ray in his lower field, wearing bib overalls and standing waist deep in young and succulent broomgrass. This is a heroic portrait — Ray's muscular forearms folded pridefully across swelling chest, a late afternoon sun playing upon a quiet smile — it reveals on home ground the pride one might have expected to see in the Awards Dinner photograph. This could be a socialist-realism poster photo, advertising

"Pride in Accomplishment" to spur younger farmers on to greater efforts for the state. But its significance is quite contrary to that notion.

"Some folks like to be ahead," Ray explains it to me as we study the scrapbook, "but prefer not to show it — that's a characteristic of the Yankee breed as a whole. None of them will admit prosperity, and yet they may be reasonably successful. Now you take Henry Dwight — up to Colrain — he was Bill Ryan's grandpa. He had only a few cows, on a sidehill, but he was away up there in production. You'd never get a hint of it to talk to that man."

The details of Raymond Totman's own history are for the most part easy to anticipate — in such and such a year he bought a new tractor, tried a new mixture when seeding down the mowing, sold so much timber, fathered children, and every day of every year, made milk. But it is the attitude, the bearing of his life, and not the string of agricultural and familiar events that so distinguishes him and accounts for his extraordinary success.

The span of his farming life is an eventful one that reaches from an era when things were done pretty much as they had been done for hundreds of years before to these modern times when nothing is done as it once was. He recollects a boyhood when Frederick was still "general purpose" farming, which is to say Frederick kept his business safe in a wild economy by having *many* purposes — he milked cows, but also raised hogs and sheep ("driving them through a covered bridge was as easy as driving them into a barn"), chickens and grain crops, made maple syrup and logged in the winter. "Butcher and grocers would come by on carts then," Raymond recalls. "The grocer's cart had a box in back full of loose doughnuts. We'd run up and hitch onto the back as he was leaving and have a go at the box — did it every time.

"During the First World War, so much was still done by

hand, but there got to be quite a shortage of manpower. I was in my junior year at the Conway high school, but starting in April I stayed home and worked, and kept up with the books as best I could on my own. Dad hired the minister to help out a few days a week. I had to drive the horses. I had learned to cuss whenever I drove the horses, and as long as I could cuss they would mind and I drove them pretty well. But I just couldn't cuss around the minister — you didn't do that. If he wasn't in sight, I'd cuss and drive pretty well, even if he could still hear. But when he was watching, he must have thought I couldn't drive very well, because I wouldn't cuss, and the horses wouldn't mind.

"I got very set on what I was doing sometimes, and had a hard time changing course. I loved to take the dog — Topsy, same one chasing Grandpa when he mowed — and go out with my dad hunting skunk. I remember one night I was begging father to come out hunting and he said, 'O.K., take the dog and go down below the next farm and then turn right downhill, and if the dog takes up barking, run back and get me.' I started out. The dog started to bark. I headed back for Dad, and got halfway home when I heard his voice in the darkness, asking, 'Where are you going, son?' 'Back to get you,' I answered, and kept right on running."

Raymond remembers his first difficult years back on the farm after he'd returned from college and was slowly supplanting his father, shifting the emphasis of production increasingly toward dairying. He remembers the first milk truck to come to the area for daily pickups ("you know the driver of that truck was a black man?"), an event that happened in the late twenties, shortly after the town first began to keep roadways free of snow all winter. He remembers the hard days of the Depression. "Shehan of Holyoke picked up the milk then — three-and-a-half cents a quart we got at the door. The week the bank folded he came along with the driver, paid us in silver. He said to me,

'Next week I may pay you in buttons, because that may be all I have left.' I said to him, 'I can use the buttons; you take the milk.' "

Raymond remembers the first tractor he owned — a steel-wheeled Fordson — and was the first in his neighborhood to make the important switch from steel to rubber tires, a switch that changed the nature of farming, as the switch from oxen to horses had a century earlier, because it improved drawing power by so much. "Rubber tires cost as much again as the whole tractor had, and Andrew Hart claimed he could outdraw me with steel wheels, so the dealer staged a draw. Andrew didn't win."

Raymond has been around during the complete transition from hand milking local stock for local consumption to machine milking genetically engineered supercows for consumption anywhere. He was always in the forefront in adopting new innovations. His milking machine, pipeline, bulk tank, baler, bale thrower (and rubber tractor tires) were all among the earliest in town. His cows have long given more milk than most other farmers' cows, in or out of town.

His days, Ray says a bit mournfully, are now spent in small chores and reading. He points to refolded *Greenfield Recorders* and old *Springfield Unions* on a table in the corner of the dining room. He says he usually has a book or two going as well. He favors best-sellers such as *Airport* and Vidal's *Burr*. He calls them "informative, but too explicit for my taste."

Summers Ray still does tractor work, and twice a day, summer and winter, in sickness and in health, following the rhythm of work he has followed all his days of milking cows he still emerges from the dining room in rubber barn boots, crosses the dooryard and "does calves," feeding them hay, a ration of grain, milk replacer, watching them for signs of ill health. The ritual takes about half an hour each time he does it. During the winter Ray fills a crate with wood from the woodshed on his way back to the dining room, lugs it up the back steps to the kitchen, and leaves it by the range for Mildred. Then it's back to home base, to the recliner by the dooryard window.

Like Saint Alphonsus Rodriguez, who, Gerard Manley Hopkins says, "Could crowd career with conquest while there went / Those years and years by of world without event," Ray Totman, for all his insularity, has come to an animating and sustaining knowledge of the world. A discipline, born perhaps of that tense yet sedate Yankee style of ambition, of wanting to be ahead in the most restrained way possible, "to be ahead but not to show it," has guided Ray's developing farmstead and character over the years.

He has applied to the problems of farming a locally grown ambition, a suppressed zeal that combines personal strength, integrity, inventiveness, alertness, even piousness. His brand of person can find no fit place in today's urban commercial life and is therefore seldom found off-farm. Ray did what he was supposed to do, with a liveliness that made it work. Other farmers caught up in the increasingly difficult conditions of Yankee farming were betrayed by their own ignorance, or their own traditions, or sloth, and their land was taken by innocent suburbanites and culpable real-estate profiteers; Ray Totman's holdings grew. While others fell prey to letting things slide as times grew harder, Totman (as sanctimonious ministers of his own era might have said) remained steadfast — which is to say he stayed home, worked hard, and was alert to every possibility of increased efficiency the new technology offered.

For children of the postwar years, brought up in a world where there is little reason to expect that virtue is what is rewarded, there is a temptation to find the archconservatism of the likes of Ray Totman naive, to suspect that because he is so very smart and warm, if he only knew a bit more about the complexity and treachery of the world, he would not be so quick to leave every man responsible for his own well-being nowadays — to suspect that he would have to admit some flaw in a society that generates whole classes born to exploitation, trained to be ill-used. Totman, of course, admits to no such thing. To the contrary, he chides me, promising that age will give me more sophistication on this question.

He was raised to hard work, and learned early that hard work pays. He admired the immense energy of his own father, who not only ran the "general purpose" farm but also worked out all the while as a teamster, hauling up high tension electrical towers with oxen as the net of power from the outside world was knit across Franklin County, and logging large tracts of land on contract. Ray always worked on through his sixteen waking hours each day. Mechanization drew his farming out of the age of drudgery. But for him each new machine has meant only that his scope as a good farmer became greater, that he could do more of what he intended to do all along. He now recalls each new machine in his life with a boyish avidity that smacks of musty issues of *Boys' Life*.

There is on the Totman farm, in Ray's house, in Lee's house, even in the cows' barn, the wholesome 'can do, sir!" atmosphere of a Horatio Alger novel. In this place the rules still do apply. Probity dominates. Success is indeed the reward for goodness. He who is the disciple of self-discipline is fate's friend here.

"He started out life on his own account," as Alger wrote of another diligent Yankee, Benjamin Franklin, "and through industry, frugality, perseverance, and a fixed determination to rise in life, he became a distinguished man in the end, and a wise man also, though his early opportunities were very limited."

Late in corn season when I came to call, Raymond answered the door enshrouded in a knee-length Harris tweed cape, bound with a braid of silk rope. He looked ashen and sent me away, "for fear you'll catch whatever damn thing I've got."

When I came back a week later he was fit again. I remarked that he seemed to have recovered.

"At my ancient age," he replied, "you never really recover — just some days you slide downhill slower than others."

I asked if that wasn't being unduly hard on himself.

"I'm failing. I've paid attention to facts for too long to fool myself. When you're well into your seventies and each day you're a little weaker than the day before it can only mean one thing."

I asked if he wouldn't see a doctor, then, because sometimes things weren't as bad as people imagined, and perhaps whatever was weakening him could be treated.

"I know what I've got. I don't want to go into any hospital." Raymond paused for a moment, then relaxed and smiled. "I'll tell you one thing for sure, though. I'm not losing any sleep over this."

I said I thought that if things were as serious as he implied, it might be difficult to face up to.

"It would have been at your age. I had work to do. Now I've done it. I've come to terms with myself. I don't have much left to do but enjoy what I see out the window." Then he sat me down and fed me tea, and after much discussion of a great many topics he pointed out the window by the recliner, out into the cornfield where his son Leland still labored, and said, "Any old farmer would like looking out and seeing his son at work."

In every rural New England county there are a few families that so clearly seem "elect" that it is easy to imagine the palpability to the citizens of Plymouth Plantation of that regal Puritan notion of inborn elitism. What is especially amazing about the last three generations of the Totman family is that they have made their farm thrive while New England agriculture on the whole has suffered sharp decline. Like horses so good they win races while carrying extra weight, a few New England farmers seem only to improve under the strain of adversity.

In these modern times it is cheaper to grow almost everything somewhere else. Less than a sixth of the farm families that were in business on the eve of World War II are still farming in Massachusetts — fewer than four thousand farms remain in the entire state. The very large agricultural equipment available now is cheaper to operate on large, unbroken fields farther west but doesn't pay on small and hilly, widely separated New England fields. The coming of milking parlors, of automatic feeding, and of high-production breeding have made efficient herd sizes far greater than can usually be justified by the low density of farmable land in New England hill country. A supply system geared to national supermarket chains favors larger and larger units of output. High energy costs, high shipping costs, and competition from the world marketplace all make it less economically feasible than it once was to grow grain in Iowa and feed it to cows in Massachusetts. As the number of farms drops, farm suppliers and food processors also go out of business.

Yankee farming today takes place in an atmosphere of siege. The banker is always near at hand. Only the sleekest, most business-minded operations can compete at all in New England. Yankee farmers who barely scrape by and who therefore think of themselves as chumps and half-foolish hangers-on nevertheless farm more efficiently than the worst farmers still surviving in favored dairy areas such as Wisconsin and upper New York State. The very finest Yankee farmers — the ones who, like Lee Totman, have combined the luck of owning some of the good farmland that is scattered about, with the wit, vigor, and application needed to manage complex modern operations well — these farmers are among the best agricultural practitioners in the United States.

Dick Mayer, Lee's neighbor and a first-rate farmer himself, once explained, "Farmers basically compete with one another to be the first ones to do things the most efficient way possible, to be ahead of the next guy in adopting the right new technology, the right new methods. Raymond, and now Lee, has always done that. They're farmers' farmers."

That shadowy autumn day, after having heard the story of Joshua, Frederick, Raymond, and the mowing machine, and having taken my leave, I step out the door into the lush late afternoon. It has sprinkled but the ground is still dry under the maples. I walk down the line of trees and head down the wide farm road into the field where Lee is finishing up. It is nearing time for evening milking.

Leland Totman is hauling a wagonload of corn uphill. He appears in the distance, an agricultural centaur, head and then tractor rising up from behind a hill. He is far down the roadway that climbs from the bottomland at the fork of the South and Deerfield rivers, up through a grandstand of pastures and mowings and cornpieces to the dooryard of the farm. Behind Lee, all along the road, are the fields that feed the cows that make the milk that Lee sells. The nearest mowing looks like a roughly overgrown lawn, rolling downhill for acres. It ends abruptly at the edge of a field of tall corn. The corn leaves have browned from a touch of early frost. The air smells sweet and thick from the day's work.

As the tractor draws nearer, Lee's resemblance to Ray seems clear. Lee is short, broad, blond, and blue-eyed, built ruggedly, molded in the same inverted trapezoidal shape as his father. He carries himself heartily. He is forty-three. He waves and flashes the conspiratorial and boyish grin. It's the grin of someone amused with what he knows.

Among other things, Lee Totman knows how to farm. He was named Massachusetts Farmer of the Year for 1977. I have heard that when the agricultural extension agents arrived to tell him of the honor he laughed and told them their award was nonsense. I have heard that his cows give more milk than those of other farmers in Franklin County, and that he would be at home in a small crowd of master farmers anywhere in the country, although he seldom goes anyplace. The tractor arrives, and Lee leaps to the ground, still grinning.

It is fitting to introduce Lee on his road, because years ago — in fact, the first time I ever heard the name of Totman — the subject of the local gossips was the roadway. What I'd heard was a tale of a "crazy young fellow down to Conway" who'd taken the family silver and squandered it on a superhighway through the farmstead.

It's a private road, used only by Ray, Lee, and Dick, the stalwart hired man, yet it does appear to be better than some of the public dirt roads in this hill country. Nearly a mile long, it is graveled on the leveler sections, and is actually tarred and ditched along the steeper grades. The road, as it turns out, is not a superhighway, but it is something special, the like of which is not to be found on any other farm in Franklin County. It represents a timely philosophy of capital investment out of step with more desultory management tactics practiced on most of the sidehill Yankee farms left in the region. When Lee first was building it, the road must have seemed to some of his neighbors as wild an investment as Joshua's purchase of a mowing machine had seemed to the neighbors' great-grandparents a century earlier.

I grin back at Lee. I comment on the road, on how unusual it is, how I've seen other farmers on lesser crosslots roads wallowing in mud when they would have preferred to be out working, and Lee says — rather sternly, grin notwithstanding — "I've been in their shoes. That's why the road's like this now."

I say I actually have expected, from the tales I've heard, that it would be a broad tarred boulevard. "People blow things up. There aren't a hundred yards of tar the whole length of it." Lee seems a bit testy hearing about these rumors. "It didn't cost me much — saved me plenty. I'm thinking of some days we spent bogged down along the old road. Now you can get through here — even during mud season. Everything that goes from field to barn or the other way passes over this road. That's a lot of tonnage and a lot of dollar value."

Lee starts up the tractor again. I chase him on foot as he drives around to the silos by the barn. I arrive as he maneuvers a silage wagon up to a silage blower. He starts the blower and the chopped corn plants — ear, leaves and stem alike, all in spoon-sized bits — hurtle up sixty feet inside a tall stovepipe, then topple down into the nearer of the two big round cement towers. The corn will ferment in the airtight silo for two weeks. It is then preserved, standing ready to feed out to cows all winter. "City people may not realize it," says Lee, "but hay isn't simply hay, and corn silage isn't simply corn silage."

That may be about as close as a real Yankee can approach to boasting — it's certainly the closest that I ever heard Lee Totman come to it. Hay may be stemmy, old and leafless, moldy, rained on, and composed of lackluster naturally seeded weed grasses. But the hay stored in Lee's barn is young, charcoal green, succulent and tests to a high leaf-to-stem ratio. It consists of the choicest planted alfalfa, ladino clover, timothy grass and red clover. It is likely to contain 15 to 20 percent protein, while the hay in lesser barns may test out to only 10 or 12 percent protein.

And the same is true of the corn now blowing up into Lee's silo. On a nearby road is a farm with a bowed wooden silo into which is grudgingly dumped silage chopped from the meagerest of corn stands, corn that has wriggled out of the damp earth intertwined with morning-glory vines, corn that has matured waist high and spindly stemmed, its leaves pale green from want of nitrogen and streaked with purple from want of phosphorus, corn mostly leaf, with the rare ear that has happened along scarcely filling out and never growing very long. But the corn Lee has hauled up the farm road behind the tractor is fare fit for regal cows, cut from fat corn plants that have tasseled out higher than a tall man can reach, that have borne one and sometimes two long full ears of squat, deep kernels, plants the bright deep green of a child's crayon drawing of a corn plant, corn harvested in the prime of "hard dent," when the sugar content and the nutrition are at their best.

The good corn, like Lee's good road, reflects his unusual attitude toward capital investment. Today it takes money to make money farming. Once in early June I saw Lee plowing under a stand of alfalfa and timothy that any other farmer in the county would have been proud to grow up and harvest. Lee said, "There didn't seem to be quite enough alfalfa in it." When he got done plowing and disking and liming and fertilizing, he reseeded the mowing with just the mixture he wanted to see growing there. If the hay is ready to cut but looks likely to be rained upon, Lee will chop it green and make grass silage out of it, rather than feed his animals rained-on hay. Few farmers in this rainy state have set themselves up with that flexibility, even though losses of many hundreds and even thousands of dollars of feed value may ride on an afternoon thundershower. When at a busy time of year some of Lee's hay did get rained on, he dried it out, baled it up and gave it away. I asked him why he didn't sell it, and he said, "I sell milk, I'm not in the hay selling business." And, when Lee culls a cow from his herd, he always sells it as beef, even though it might be brought back to health by a patient farmer, and then would almost certainly be a finer animal than most cows in the region. "I'm not in the breeding business," Lee says; "I'm in the milking business. I can't see saddling some other farmer with my problems."

Lee does what is necessary to grow first-rate corn and hay. New England fields tend to be too acid for the job, so every year he spreads two or three tons of ground limestone on each acre of tilled ground. The limestone is followed by lavish applications of cow manure and of bagged fertilizer as well. Lee grows corn on land that grew corn last year, but not the year before that, land which, after the previous crop had been harvested, was seeded down with a winter cover crop of ryegrass to be plowed under just before spring planting.

Each job of fieldwork is a tractor trip back and forth along miles of cornrows and haylands. It costs time and fuel and equipment wear, and every pound of lime and fertilizer and seed added to the ground costs still more.

All these practices pay back in increased feed value of the crops, but when most local farmers — who hold little cash reserve — envision the length of time before their improved feed crops are planted, harvested, stored, fed, eaten, and finally make more milk, and therefore more money, their old-fashioned thrifty instincts frequently lead to their doing the best they can with field conditions as they are. They keep their money in their pockets. When local farmers feel cramped for funds to pay the mortgage, the first place to cut is on investment in field upkeep. Fields in good shape are like stored money, and negligence in tillage practices — a bit less lime, letting the hay seeding stand one more year, hedging on fertilizer application — saves scarce cash reserves.

In the light of this kind of thinking, Lee's tillage practices are especially remarkable. The pressure upon him is relentless. The land, used in the context of the belligerent and harsh commercial structure of modern American farming, teaches the most compelling and relentless of work ethics: free lunch today; no supper tomorrow.

The silage wagon is empty now, and Lee draws it back through the dooryard and parks it in the machine shed, shouting to me over his shoulder as he drives past that I should step down to his house with him and "grab a bite to eat." "But there's a lot to do first," he adds, shutting off the tractor, "a few chores to do up here to start with. The cows eat before we do."

We walk into the barn. It is long, red, and wooden, as barns are meant to be (although an alarming proliferation of modern barns seem to be aquamarine, and stamped out of steel by the thousands in some midwestern factory). It's not, however, what tour guides bill as a traditional New England barn.

It is only a single story high — hay is kept in a shed behind the barn, rather than in a loft on top of it — with the two fat cement silos rising against it. The barn's "great hall" is called the "loafing shed." Not a cow in sight in the farthest corners of the hanger-like room. It is cool in here and it smells of corn and sawdust, manure and leather. It is spotless. The walls are whitewashed, the floor is scraped clean and new sawdust bedding added after the last cow wandered out to pasture this morning.

The cows spend these last warm days out in pasture, and will do so for another two weeks, until the silo is sealed and the new harvest of chopped corn within has had its fortnight to ferment. When the corn is ready — about the last week in October — the fifty-five or so milking mothers, their young and growing progeny, one scrub bull used to "clean up" problem breeders, and one Holstein-Angus steer, half-grown and destined for the family freezer, will all be herded inside for the winter — eighty-odd animals in all.

Then Lee will divide the milking herd into two lots. Heavy milkers — usually those recently with calf and therefore early on in the ten-month milking cycle — will go in one half of the loafing shed, and the lighter milkers — later in the cycle — will go on the other side of a cable and chain fence. The heavy milkers have access to a trough fastened to a sidewall of the barn, where Lee feeds them extra grain to sustain their production. A long cement feed bunk runs up the center, like the mid-kitchen lunch counter of a suburban tract home. Heavy and light milking groups eat from opposite sides of it.

We walk through the loafing shed. Lee fills the hayracks on the walls with his beautiful hay, glances into the trays offering mineral supplement, then checks the salt licks strewn about the central trough. He opens the door leading to the base of the further silo, reaches in, and closes a big switch. A motor whines, chains clank, belts slap into motion, and far overhead in the silo, an automatic unloader begins to work its way around the top of the tower of silage. A conveyor delivers a stream of feed

— nearly a ton — to a hopper with a spout. It scoots along, suspended from an overhead rail, dishing out supper all along the central feed bunk.

"What I hope every afternoon is that the commotion of the silo unloader will begin to attract the cows from the barnyard. They are pretty well trained to know that food is being served up here about now. Some farmers still feed silage by the wheelbarrow load. Takes them an hour to do it and they have to holler after the cows as well. That big motor up in the silo usually makes enough noise to interest these girls. They are anxious to be milked. It usually works." As he talks, a few cows do indeed amble up to the open door of the loafing shed. They bridle and hold back when they see me.

Cows are upset by anything that isn't just as it was yesterday. After a moment, though, they step on in and begin to feed. Lee watches intently as the cows gather, taking in details of gait, vigor, aggressiveness, hunger, thirst, changes in the strong pecking order of the herd, cows riding other cows, which they do when they're in heat. His is a beautiful herd, a herd of Cadillacs, large, clean cows, with coats shining and what cattle judges call "dairyness" — the undefinable look of an animal that means business at milking time — emanating from every beast.

Lee wanders in through the herd. He talks now and then in an odd voice, a flat monotonic sound he makes while tucking his chin into his chest. It seems to come from deep in the throat, as if he is imitating a voice heard over the telephone. He is greeting his favorite cows.

"Hey, *bos, bos, bos*, there, girl," he is muttering. "Don't you look fine, girl."

Bel canto an aside to me: "Just calved."

Sotto voce to the cow: "You're a pretty good one. Hey, Wendy; hi, Dinah girl, there."

"Why do you say, 'hey *bos, bos, bos*'?" I ask Lee as he closes the barnyard gate behind the last cow in.

"I've always called cows that way," he says.

"Who'd you learn it from?"

"Learned it from my dad, I guess." Lee has fully recovered the use of his own pleasant baritone voice.

"Your dad — he must have learned it from his dad, too?"

"Never seen it in a book, I'm sure." *How to Call a Cow*, I think to myself, or perhaps *Cow Calling Self-Taught*. Lee goes on, "All the farmers around here call their cows by saying *bos, bos, bos*. Why?"

"You know the Latin name for the cow?" I ask.

"Don't happen to, offhand." He's remembered his grin now.

"*Bōs. Bōs taurus typicus prīmigenius*."

"I guess my dad's grandpa's grandpa must have heard the word from Julius Caesar, then. Never knew that."

While we are chatting, we walk back through the loafing shed to a part of the barn called — from the time when it was a separate structure by the road with a cool spring running through it — the "milk house." We enter through a dark hallway, noisy with the roar of the compressor that cools a stainless steel milk tank the size of a compact car. Milk is stored there, and picked up by a big tank truck once every other day, by which time Lee will have taken enough milk from his cows to fill between three and four thousand quart cartons at the processing plant.

Inside the milk house, in racks all about the room, rest spotless pails, hoses, lengths of stainless and plastic pipe, milking machines, all washed in nearly boiling water, disinfected, dried and ready to be reassembled for the coming milking. On the wall are two separate sinks as per state regulation — one for milking equipment, one for the sullied hands of the farmer. Lee tells the tale of one irate fellow who didn't feel he needed a separate sink for his hands. He read the law through and discovered that while he was required to have the second sink, its height above floor level was not stipulated. "He put it up about ten feet in the air, and I guess he enjoyed the inspector's next visit pretty well."

Lee goes to work now, doing what he does twice a day. He starts reassembling the milking machines. Most farmers in the area still call their milking equipment "milking dishes" from the days a hundred years back when cream for butter making was settled out in shallow pans right in the milk house. Nowadays, whole milk is shipped to large centralized processors, and the remaining dishes aren't pans at all, but piping, glass measuring jugs, yards of stainless steel tubing leading directly from cow to bulk cooling tank, and claws of rubber milking "inflations" dangling from fist-sized transparent pods — floppy juggling mouths of soft rubber that enfold the four teats of each cow and draw milk from her in gentle wet pulses engineered and improved in a hundred subtle ways to adore the udder and more pleasingly resemble the mouth of a feeding calf.

Lee carries the assembled milking claws into the parlor. This is the payoff. Everything's ready and waiting. It's a small room, walled with ceramic glazed tile. "We got a bunch of seconds — new tile costs an awful lot," Lee says when he sees me taking in the parlor construction. In its center is a waist-deep pit surrounded by six stalls. The cows walk into these stalls, are milked out, then pass back into the loafing shed. Now the parlor is clean, empty, dry and ready to go. Lee swings open a pair of heavy entrance doors by tugging down on a rope that dangles from pulleys over the far end of his working pit. The race is on. The lead cows are in a hurry. They have heard the sound of the sweet grain supplement falling into tin troughs — at least a dozen pounds of grain per milking per cow. It sounds like hail on a tin roof. The troughs are mounted at the front end of each of the parlor stalls, three feed bowls heading up three stalls on either side of the central pit. As the cows charge in, Lee names them, using the same telephone voice he used to greet them earlier. "Hey, Terri; hey, Judith; hey, Karen," he intones as the huge animals lurch into place and stop suddenly, each bushel-sized head dunked into its own private world of luxury. It's as if each has won a supermarket giveaway contest and is immersed in all the grain she can devour in seven minutes of hard eating.

Lee stands waist-deep in the pit; the stalls are arranged so that without stooping or stretching he can work at chest height on the udders of each cow in turn. Now he starts on the first of the three cows to his right, washing the animal's full bag with warm water and a sponge. It takes about a minute for the cow to "let down" her milk — the washing is a stimulus to which a first-calf heifer is conditioned as soon as she enters the milking string for the first time. Washing causes the muscles of the udder to relax, a chemical reaction triggered by the enzyme oxytosin, and the milk, which has been building up in the spongy tissue of the udder, descends freely into the streak canals of the teat, whence it is drawn out.

By the time Lee has washed the third cow a minute is up. Returning to the beginning of the line, he attaches the first milking claw, then does the same with the second and third cows. The milk spatters and dances as it is sucked in spurts through the plastic tubing connecting claw with collection bottle. This is harvest time. Lee moves around to the three cows waiting and eating grain frantically on the opposite side of the central pit. "Hey, *bos*; hey, Sarah, there's a fine one; hey, Esther; hey, Tracey." While he repeats the same procedures the first three cows stand, plumbed into the machines, into the barn, milking out.

The milk from Lee's trouble-free animals moves from cow to collection bottle to bulk cooler, and on board a tank truck, to the processing plant untouched by persons. The milk taken in Lee's gleaming and aseptic milking parlor always has a low bacteria count — there's no question about it. Cleanliness, for all its proximity to the deity, is an accomplishment of procedure, as workaday as loading up the dishwasher after supper. The rubber and stainless steel calves that draw milk from bovine mothers enter each milking forbiddingly disinfected. The machines do, however, still make a sound like live calves suckling; they pulse with the slow and imperious rhythm of living things. It's a steadying and hypnotizing swish that breathily repeats itself with about the frequency of a healthily beating heart.

The first milking machine first hung from a cow in the 1830s. The grossest refinements were thought up and tried out before the machine became practical. Inventors once attempted to extract milk by inserting hollow ivory tubes into the cow's teat ends, far enough up to distend the muscles restraining milk flow. Milk flowed, but infections followed. A later series of patents toyed with mechanical rollers that repeatedly wrung the teats of hapless cows. The climax invention substituted vacuum for mechanical manipulation, just before the turn of the century.

At first, though, farmers saw little advantage to mechanical milkers — the new devices were hard to wash, costly, and, because labor was relatively cheap, the money saved by not milking a small herd by hand was minimal. Mechanical milkers came into wider demand only when the First World War suddenly took many experienced hands away from the chores. Still, only 11 percent of eastern farmers had automatic milkers in 1919; 20 percent had them just ten years later. Raymond Totman brought his first mechanical milker to the farm in 1926. Its appearance relates to his own succession to power on the farm. It happened a year or so after the morning at the woodpile when Raymond and his father finally got around to negotiating the terms of their partnership. Raymond was coming into his own as farmer-in-charge. His father was still doing evening milking, but Raymond wished to take over the job himself.

"Father was," Raymond told me, "a wet-hands milker — some of the old hand-milkers couldn't do a thing until they'd dipped their dirty hands into the milk and made 'em nice and slippery. Now that might have been very good for gentle milking, but it was bad for the milk and bad for the cows — gave the milk a high bacteria count, and gave the cows mastitis. But you'd be surprised how many of the old-timers did it. Lucky for me, in a way, Frederick had arthritis pretty bad too, which had weakened his hands by then. A dairy started offering a premium for clean milk, and I wanted that money, so I set up with DeLaval's gas engine milker in nineteen twenty-six — used my dad's arthritis to ease him out of it gently."

Raymond milked cows all of his farming career by hauling "bucket milking machines" from cowside to cowside and plugging them into a vacuum line. The cows stood in place to be milked, clamped in lines of stanchions like offending Puritans gripped in stocks. The idea that the farmer might stay put with a battery of milking machines by his side while cows passed by to be milked — the innovation represented by the milking parlor — had been conceived when Raymond was still a boy. But save for a few showplace dairies, farmers found that it didn't yet pay to construct parlors. It wasn't until 1970 that Lee decided to make the changeover.

By then the farm was well established, and firmly capitalized — there was money to spend on it, and tax benefits to encourage doing so. The parlor allowed him to milk about a third more cows in the same length of time and in greater comfort than before. It saves the knees. Retired dairy farmers all have bad knees. The increased efficiency, gained in trade for a goodly capital investment, had become crucial to survival in the stressed economy of the 1970s. And a new technological insight in the field of barn design had recently made the parlor system of cow handling still more attractive.

Raymond's cows, in stanchions, delivered their manure to a gutter that ran behind the line of animals. A mechanical chain scraper shoved the manure out of the barn. It beat shoveling; Ray liked it because he had put in years of shoveling cow manure. He had thought of installing a parlor himself but had decided against it because of the problem of manure disposal. Once rid of the stanchions, how did one keep the barn tidy? The problem remained into the mid-fifties, and was finally solved by Ivan W. Bigelow, a dairy barn engineer in the service of the New York State Extension Service, who invented something called "pen stabling." Cows were fed in one shed, which was scraped clean daily. They were housed in a bigger shed, which remained unshoveled but was heavily "bedded" every day — sawdust was thrown about the floor in great quantities. It sopped things up, and matted down, and the cows lived higher and higher off the barn floor as winter progressed. Finally, when the herd dwelt near the roof, a stable cleaning took place that would have taxed Hercules. Then things started off again at floor level. Parlors began to appear as farmers with larger herds switched to this system. It is still in use on many farms, and has its champions.

Another system, far easier to manage, has generally supplanted loose pack housing. A farmer from Snohomish County, Washington, Adolph Oien, is the father of the "free stall barn." "Late in nineteen-sixty," as Mr. Oien related the story to me, "I was handling about forty cows and needed to find time so I could handle more. It took so much bedding and so much time to handle manure with loose pack housing. I can tell you what gave me the idea — I was using a feed alley to feed, and there was a space next to the bedding area, always had bedding kicked over. It was four feet long, and nine feet wide, and there was always a cow laying down in there, and she was always clean, except it was soiled at the far end, which made extra work for me. I thought one day, what if I turn it lengthwise, make it four foot wide, eight foot long, and arrange a bunch of them in a row? I was so sure of myself I didn't even build one first in the old barn. I built a fifty-cow shed that way right off the bat. It was the day before Thanksgiving, nineteen sixty, that I opened the door to cows. Within fifteen minutes they started to lie down in the stalls, and the stalls stayed clean. Within a few months, people drove me crazy. I couldn't do any work. I had people from all over — Germany, Japan, fellow from England. It calmed down within a year, because there were so many others built. The fall after I built it, they were sold at fairs by implement dealers. I didn't get a nickel from it."

Behind Mr. Oien was Mrs. Oien, who, according to her own recollection, played an instrumental part in the decision to build the new-style barn.

"He came in and said, 'I ought to put up little walls all over the barn.' He kept saying it, and finally I said to him, 'Why don't you try it?' So he did. If we could have a penny for every stall built since then, we could retire. As it is, we're still out there, farming."

What Mr. Oien conceived and Mrs. Oien urged into existence is one of those truly elegant constructions that is obvious only after it's been made. He filled his barn with rows of soft bedded spots separated by small sturdy fences that kept cows facing forward and lying down after they entered. Today, most barns of larger dairies, like Lee Totman's, provide free stall housing. Cows rest in the bedded stalls, and wander out into the concrete aisles to eliminate and to feed. In a well-designed free stall barn, a farmer on a tractor with a scraper blade can clear the aisles in a very few minutes a day. Lee Totman's barn chores are minimal. He feeds by switching on the automatic feeder and he cleans with the tractor. The cows eat when they please, sleep when they please, and, quite remarkably, they line up on their own twice a day for the chance to walk into Lee's parlor. Once there, they luxuriate, seemingly enjoying the event. They linger for a last bite of grain even after they are milked out.

Milking time for Lee Totman is routine, predetermined, and as telling as the counting of ballots with the polls already closed. It is therefore a time for Lee to brood upon the overabundance of querulous information about each animal, to plot still further tampering with the life of the herd. And when that's been done for long enough, and then for some time longer, there is time left twice a day for thinking about life and the state of things in the world. It's a time of solitude and repetitiveness, a meditation on a milky mantra, a compulsory and twice-daily trafficking with motherhood in its mammary, leathery quintessence.

As each of the first three cows finishes giving milk, Lee removes its milking machine and reattaches the machine to a cow, washed and waiting, on the opposite side of the central pit. When all three machines have been reattached, he sprays the finished cows' teats with disinfectant, then opens a gate on that side of the pit and mutters, "Go on, girls." The trio swaggers out an exit lane back to the loafing barn. Lee admits three unmilked cows to the emptied stalls, washes their udders, turns back and detaches the three milking machines from the cows now finishing on the other side, puts them onto the new cows, turns again and sprays the new dozen milked-out teats, lets out this set of finished cows, lets in three more cows. It's allemande right and left for the next hour and a half as milk, at thirteen dollars the hundredweight, sloshes into the bulk tank.

As the cows come past, as he works in well-set routine, he watches the cows. Perii is in heat; she is fretful. Jake is limping and wants her hoofs trimmed. Sarah is an unusually friendly beast. She trots in and lowers her head as she tries to nuzzle Lee. It is quiet in a sense. That is to say, there is a terrible din — compressors, pumps, and that calf-sucking sound of a dozen teats simultaneously taxed. Lee says that for him this is the comforting noise of routine, the silence of a clockmaker's shop, the sound that says nothing unusual is going on. Occasionally Lee mutters endearments and an occasional observation ("Good girl," or, "Keep that tail over there, you") as each comes in, stays a few minutes, and moves on. He is using that telephone voice again, naming the cows as he handles them — many have playful names. There's "Red Ric," "To Rose," "Rabbit," "Auntie Boo," "Tippy Tu," "Mighty," "Bighty," and

"Oozy," who prefers lying in the messy alleyways and not in the clean-bedded free stalls, but was named as a calf before she showed this nasty trait. There is a whole bovine clan called "Pig Tail," "Pig Head," "Piggyback," "Pig Pug," and "Piggly Wiggly." Lee says to me, "I raise pigs as well as cows."

Leland is alone and he forgets himself, lost in the intricate and familiar chore at hand. More cows come in, stand for milking, are urged out. New cows enter. His is a laborious, exacting and tedious stewardship, rewarding in satisfaction and increasing numbers on net-worth statements. The consuming nature of the profession makes dairy farmers a breed apart — a people who dwell amidst a passage of events crucial to their own well-beings, but so intricate, picayune, private and absorbing as to be incomprehensible to most outsiders. Lee Totman's vocation insulates him from all but a few close neighbors and family members in the same trade. He entertains himself as he works.

As the second hour of milking starts, Ray Totman comes into the parlor. He nods a greeting down into the pit, toward his son. His entrance has been orchestrated and he has arrived just as "Mimi" is being milked. She is a fresh cow, has calved only a day ago, and is still giving thick colostrum milk, which can't be marketed but contains antibodies newborn calves need. Ray places a bucket under a collecting jar and drains the milk off into it. The young calves live just behind the parlor. He leaves, lugging two full pails of milk. Lee has barely acknowledged the old man.

Ten minutes later, pails empty and rinsed, he comes back. He places one on the steps leading into the working pit and sits on it, ready for a chat. Lee finally nods back to him. Raymond watches Lee for a few minutes.

"I hear Melnick's been having more trouble. That fellow who works for him was saying. More stuff in the corn," Ray says.

"What was it this time?" asks Lee.

"Railroad car springs. Sawed into rings, then slipped over the ears of corn, then the husks of the corn tied back over them."

"My God, but whoever's doing that is getting professional," Lee says. "That'll play havoc with his corn-chopper. Did any get into his machine?"

"A few. Says the machine jumped right into the air when it went in. Caused a couple of thousand dollars' worth of damage. They have a big new chopper — three-row."

"Vandals in the corn patch." Lee shakes his head. "Imagine, trouble like that way out here. I'd like to catch one of them. Wouldn't need to call the police."

"Didn't used to have to worry about that," says Ray. He reflects for a few minutes.

Lee goes on milking.

Finally Ray continues, "Never knew a case like this that wasn't a matter of personal grudge — I'd say someone knows him and knows farming. Even knows which land is his, in among everybody else's rented pieces down along the river-bottom."

"Did you hear the National Guard was there — walking the cornrows with metal detectors?" Lee asks.

Lee continues the intricate and well-rehearsed motions of milking. Raymond must see so much in those motions without even being aware of it, having milked for half a century himself.

"That heifer's come around," Ray says. Lee nods. More silence.

"Bearing's beginning to sound pretty rough — ought to last one more day of corn chopping, though. Should be through for the year this time tomorrow," Lee offers.

"Ground's going to soften up some this weekend — rain due about then," Ray says. He sits for another few minutes, taking in the scene, then leaves without any word of parting — he just slips out. Lee milks nine more cows, cleans up. Leaving the barn at 9:20 P.M., nearly sixteen hours after he began his working day, he walks the quarter mile back down the hill past his parents' now darkened house to his own. He takes himself a bowl of maple-nut ice cream, and offers me one as well. He turns on the TV and soon falls asleep in front of the Monday night football game. I let myself out. He dreams (he tells me later) only that it is tomorrow, and that he is out chopping corn again.

Lee Totman's waking dreams are not restricted to views of tomorrow's working day. Like most other farmers, he longs to be the only one in town to milk cows twice as fast as cows are usually milked. That there's no hurrying the cow at milking is one of the great stubborn givens of dairy farming — the sort of characteristic behavior that once caused Ray Totman to exclaim, "Cows are so — so *cowish!*"

For the past century dairy cows have been selected so strenuously for milk production that their mammary systems seem no longer to be in harmony with the rest of the animal. They are overdeveloped, and from Lee's point of view, oversensitive, always stressed. Well-designed and well-maintained milking equipment is, for Lee, the difference between a profitable herd and one whose milk is contaminated with mastitis.

There is an elegant and almost "political" sensitivity manifest in cows' tendency to develop mastitis. Because barns, animals, feed, labor and operating capital all cost so much nowadays, and because farmers compete with each other to make as much milk as they can as cheaply as possible, dairy cows are fed well and pushed hard. They are always in a delicate condition.

Lee's herd always includes a cow or two suffering antibiotic treatments for mastitis, or whose subclinical mastitis is controlled by careful handling. If this weren't the case, it would be a sign that the herd wasn't commercially viable, that it wasn't being pushed hard enough, wasn't bred highly enough for modern times. Too much mastitis, of course, is ruinous. But, in a curious way, so is too little. The fine art of capitalism, cows teach us, includes the art of maintaining complex biological-mechanical-financial systems in a state of sleekly managed slight debility. It does not seem to be enough to stop just before this point is reached, either. The press of commercial life forces Lee to continue striving for greater productivity until slightly after stress becomes manifest. It is a sign that everything is operating in a commercially acceptable way.

Faster or stronger vacuum pulses do not shorten cows' seven- or eight-minute milking interval, but only result in more sick cows. Modern milking parlors are so expensive, however, that farmers persist in their daydreams. In theory, if the length of time each cow has to stand in place could be halved, a parlor could handle twice as many animals for the same capital investment. To date the beast has been victorious, reacting — rather sullenly — with prohibitively increased rates of infection when milked "unnaturally." The beasts' victory may be a victory for New England farmers as well. In New England, if the handling capacity of existing parlors were to double, the result would not be that twice as many cows would then be milked. Nor that farmers would enjoy twice the prosperity.

Unlike Lee Totman, most Yankee farmers do farm widely scattered holdings — simply because in the rugged terrain of Yankeedom, there are few places outside the Connecticut River valley where there are many acres of tillable ground of a piece. With the coming of each technological improvement that increases the size of the herd a farmer can carry, New England farmers become still less able to compete with farmers in western New York State, or in Wisconsin, where large fields near the barn are more common, and where climate and geography permit grain to be grown and combined on the farm, rather than purchased. The industrialization of agriculture favors Wisconsin.

As a result, about a third of New England farmland has gone out of production just since 1960. The human dimension of this loss is particularly ugly. The most traditional farmers find themselves increasingly unable to cope with the demands of modern production. All the while thinking badly of themselves, they are forced to auction off stock and equipment and find what work they can.

Dairy farming is still among the least successfully industrialized sectors of the agricultural economy, and that has probably saved New England dairying from still more disastrous setbacks. This is not to say that the farmer who milks cows does more handwork and less machinework than grain farmers or cannery tomato farmers or hog farmers, because nowadays any farmer's daily routine consists chiefly of going from machine to machine to machine — in the case of Lee Totman, from milking machine to manure spreader to mower to baler to corn-chopper to automatic silo unloader and back to milking machine again. The average dairy farmer now spends fewer than sixty hours of work a cow each year — that's down from one hundred and thirty hours just twenty years ago. He gets a hundredweight of milk for three quarters of an hour of working time, down from two hours and a quarter twenty years ago.

Yet the reason milking is the least industrialized of all harvests is that a cow has to be harvested twice a day while a wheat field has to be harvested only once a crop. This spells an enormous difference in productivity. Milking time is still the limiting factor in dairy efficiency.

Someday perhaps genetic engineers will design a cow with an enormous udder, one that will store fifteen thousand pounds of milk, so that a cow too can be harvested once a year and compete more evenly with a wheat field. For the foreseeable future, however, the pressure on farmers to increase capital-intensiveness — to stay modern in order to stay in farming at all — is chiefly pressure to adopt mechanical innovations. And now, the pressure of mechanical innovation focuses upon milking methods.

By some inexorable working out of the laws of economics, milking time on the family farm seems to be about an hour and a half. It was an hour and a half when that was time for Lee's grandfather to milk out ten cows by hand; it was an hour and a half when that was time enough for Ray Totman, working three portable milking machines in a stanchion barn, to milk out a herd of forty-eight animals. It it about the time it takes Lee Totman, working in a double-three herringbone parlor, to milk out a herd of about fifty-five.

A new invention is being marketed, however, which, like most of the latest farm improvements, doesn't lessen the drudgery of chores — it simply reduces labor requirements per unit of production. It threatens to decrease drastically the number of dairy farmers, especially dairy farmers in New England, where operations are least suited to the new tool. Agricultural scientists have succeeded once again in reducing the number of seconds per day a dairyman must spend to extract the milk each animal continues to offer up.

The new invention, replete with minicomputer and space-age sensors, is an automatic milking machine taker-offer. The complete package, with extras, removes the milker automatically, then opens the exit gates allowing the milked-out cow to leave the parlor, opens the entrance gate to admit a new cow, and turns on and off a grain delivery auger that meters out the new cow her correct ration. It translates into a very small savings in time per cow per milking — about twenty seconds a farmer doesn't have to spend doing these chores. The result is that the farmer adopting this new technology — at the moment a costly new technology — spends less than half a minute with each cow and has free time on his hands. Farmers are among the last of us still believing that idle hands do the devil's work, and rather than daydreaming as other sorts of workers have done when freed from chores by modern technology, farmers are likely to expand the size of their herds and the size of their parlors, to milk all the cows they can in the hour or two they can take twice a day for milking time. Preliminary results show that state-of-the-art dairy farmers equipped with automatic take-off milking parlors milk up to twice as many cows — up to eighty cows per person each hour — as farmers going it the old way.

This puts pressure on all farmers to stay up with their neighbors, and the day will probably come soon when a farmer who fails to adopt this innovation will be unable to compete. It may even be that, faced finally with the need to increase herd size to a level that is simply beyond the limits of geographical practicality in New England, the surviving New England dairy farmers will follow the already well-established trend of simply dropping out. It is no longer a question, as it was earlier in the region's agricultural history, of simply switching to a more labor- or capital-intensive crop and going right on farming. Except for a few specialty items, milk is the most intensively farmed crop. Already most of New England has reverted to woodland — simply because there was in fact no good answer for the problem of how to stay in farming.

Somewhere between Conway and Colrain, there dwells a fabled and untroubled fellow who milks his cows at noon and midnight, and has for years. His scandalous working hours have cast him in the image of lionized libertine of the local dairy world, a vision of life-as-we-know-it, but with life's dear confining corset come a few hooks unlaced. He seems to other poor hill farmers to have trod a first dangerous step into dairyland utopia. Ah, to sleep late in the morning! And "late" meaning late — ten, eleven — not just "late" meaning six-thirty, seven. Ah, to stay up late into the night! To break stride with the neighbors!

One might hardly know where to leave off, once the unlacing's underway. But the lowdown, the sad truth about the fabled fellow, is that he's just plain conservative and ornery —

a Coolidge Republican, isolated, independent, and with a penchant for flapper-era dairy science — which makes him eccentric, but in no danger whatsoever of going too far. He follows the best advice of an extension agent who straggled up his road late in the thirties peddling the gospel of electricity to pokers-along as yet unelectrified. Fabled Farmer and the agent got to visiting, the story goes, and it turned out that the agent knew about other gospels as well. His gospel of milking times had it that the right-minded farmer hustles his herd into the barn and starts milking on the button, every twelve hours, summertime and wintertime, in stormy and clear weather both, putting in precise diurnal stanchionside reappearances, circling the barn like a burgher manikin rotating about on some ancient German town-square clock. It was the gospel of the same era that had the Western World's parents feeding baby by the stopwatch. On your mark, get set, milking time. And this fabled farmer followed the gospel. It's just that, over the years, starting hour crept forward a smidgen at a time. Soon, noon and midnight.

Lee Totman milks in uneven intervals. He usually starts about six in the morning, and he usually starts again summer evenings at about four in the afternoon, dividing his long working day into sections of about ten and fourteen hours. Sometimes he's an hour late. Usually he's on time "more or less." In the winter, evening milking is always an hour later. He's followed this plan for years — says the herd makes up in increased morning production what it loses in diminished evening production. Nowadays, extension agents who visit farms have come to agree with him — it's official. Farmers interested in selling breeding stock have known for centuries that cows milked late yield more, and they used to milk many hours early the night before an official measurement. "Got so it was suicide not to cheat right along with the best of them," one old-timer recalls. Nowadays milk-yield testing is on the up-and-up —

surprise first visits, testing continuing for two consecutive milkings. No horse trading among farmers any more. Nowadays everything's right in Lee Totman's world.

What is generally true of farmers' sons who have stayed home to farm — and it has been true of Leland Totman — is that they blossom slowly. The son finds when he starts working, that father is — as Ray was — at the height of his finesse, at the stage of his career when he finally has generated the capital to carry through dreams for improving the farm that he has harbored since succeeding his own father. And son is suddenly around full time — an extra hand. Big building projects start up. New barns get built, with silos added on next to them, herd size expands, brushy land gets cleared back.

A boy at home and in the barn, at his father's side and at his father's beck and call, is the stuff of dads' dreams. But a son, in such a situation, is not likely to dare draw the boundaries between parent and child that, now that agriculture is practiced by so few in New England, have come to be expected of young men. Farm boys keep their own counsel and appear, to outsiders at least, to be shy and bland into their forties, when the crustiness, the humor, the slowly ripened and long-constrained character finally becomes visible to the outside world.

Lee Totman is the eldest of four children, the only one to stay near home, and the one who most strongly shares his father's Republican views. One sister, Gail, a computer programmer by training, is now the president of a suburban YMCA and wife of a business executive. A second sister, Barbette, married to a college professor in Baltimore, is a community activist in the most liberal of causes. Lee's younger brother, Conrad, who shares his Baltimore sister's values if not her vociferousness in matters political, may be the "star in the family crown." Connie went off to the state college, then into the army, then to Harvard, was awarded a Fulbright Fellowship, a Wilson Fellowship, spent some years in Japan, where he married a woman named Michiko. He took his doctorate in Japanese history, and now teaches the subject at Northwestern University, in Evanston. He's the author of *Politics in the Tokugawa Bakufu, 1600–1843*, which was published by Harvard University Press in 1967. His mother has claimed the page proof of his listing in *Who's Who* — it's pasted up in the family scrapbook just a few pages after the photograph of Raymond at his Awards Dinner.

To be a very good farmer is an odd position, isolated, and it places Lee apart from the working community and at odds with less successful farmers in the county. It is a job that will never win him a listing next to Connie's in *Who's Who*. That seems to be all right with Lee.

"In high school," Ray told me once, "Leland was elected to the honor society. But you know what? He wouldn't even go to receive his award. I'd hoped he'd go on to ag school. He didn't want to, though, and to try and make him go wouldn't have accomplished anything — not with him."

"When I was in high school," Lee recalls, "I was interested in fooling around. Nowadays kids are too wild. We were wild, too, but in a friendly sort of way. We did pranks when we were restless. Today kids take drugs and drink instead." His grin turns absolutely gleeful and boyish as he recalls the particulars.

"I wasn't usually the leader, more like the one urging him on. Now, when Levi Davenport was in school, he was a real clown — he's a reverend now, up to Saint Johnsbury, Vermont. But then he had this trick — he'd make faces at assembly whenever the principal's back was turned. The kids would roar, and the principal would be pretty puzzled. I was always the one to get him going.

"There was a boy in high school — Fatso Horelick — got a new car. Fanciest one any kid around had. And there was a big dance. Fatso took a new girlfriend, was all set to impress her. Guess he must have done, because they left the dance early, got into his car. But they didn't go anywhere. We'd put a block under the rear axle — had one wheel just lifted off the ground — not so's you'd notice. We were hiding behind other cars. He started up — stopped to smooch a bit (we watched that too), then he tried to back out. After a bit he got out, walked all around the car, scratched his head, got back in, tried again, got out again. Then I guess he heard one of us laughing.

"Halloween was the biggest time for moving stuff — wagons on barn roofs, stuff where it shouldn't be. We coasted a man's new car downhill and hid it behind the library, right in the center of Conway. Had folks looking for it all day.

"On the Fourth of July we'd put carbide and a bit of water into a forty-quart milk can, poke a hole in the side of it, and put in a firecracker on a long fuse. One night, on the recreation field, we set one of those up and it didn't blow. Just then a drunk wandered up saying to us, 'Gee, I did this when I was a kid.' Right away he's lit a match by the touchhole to see what's wrong. We scattered but the thing didn't go off.

"We'd blow the ends out of a fifty-five-gallon drum off the Stillwater Bridge, and then light one over at the other end of town so the constable would be chasing back and forth. We had quite a time. We ran together.

"Before the war they still had minstrel shows, and we went to those together once a year. They'd come right into Conway; we didn't even have to go down to Greenfield. Four guys would paint themselves black and then tell jokes. There was a barbershop quartet also, and a banjo.

"There were dances too. I was never much for dances, but I did meet my wife at one. There was a guy drove milk truck, used to call a dance over to Ashfield. We'd go, look for girls. My wife was down from Colrain to one dance, and we started going around together.

"Nowadays, I'm too busy to fool around, dance. Haven't been to a movie in twelve years. When I left high school and started in farming full time — it was nineteen forty-nine — I hadn't seen any of the world. I hadn't been away from the farm much then, and I suppose I still haven't been. Then I just wanted my week's pay and time to raise hell — didn't even think about a partnership with my dad then. I just accept things as they come along, I guess. Over the years I grew into it."

Lee did eventually end up in partnership with his father, but it took him over a decade working as a salaried hand to get around to it. As Ray recalls it, the events evolved like this: "In nineteen sixty Lee said he wanted to change roles on the farm. He'd been working for wages since he came in, in nineteen forty-nine. He lived here in this house until he married — that would be nineteen fifty-five — then moved on down the hill to the next house. I'd talked to Lee about it, but he'd never wanted a partnership. But then in nineteen sixty he was ready for a change, and I was, too. We got some samples of agreements between fathers and sons in the same farming business, and we took what we liked from several, simplified them down, and signed up.

"For me, it was more or less a case of put-your-house-in-order. That was the year I was getting sick with cancer — I was lucky enough to survive — and Lee took over completely. He'd also decided he wanted to switch from stanchion barn to loose housing and milking parlor, and I told him he didn't want to put money into real estate he didn't own. I was ready, he was ready, and I sold him the farm.

"The deal on the sale was a demand note with a set rate of interest and no principal payments. That way I have an assured income, and the ownership has slowly shifted. I don't feel like it's my farm I've sold him, either, like I would say, selling a cow to a dealer. This farm is a thing with a life of its own — it's been my well-being, it's Lee's well-being too. People say I did it all for Lee. They don't know what goes on here or they would think otherwise. I think people are awfully jealous of success, hard work, good fortune."

Lee is quite different from his father, more abrupt, direct, aggressive, more aloof and mocking, especially when on the defensive or in public or with new people. He says "No" and "Yes," instead of "I don't think so" and "I do indeed think so," like good sidehill farmers ought to say. His head-forward bluntness has raised eyebrows at meetings of local farmers' organizations every now and then, where he has demonstrated impatience with tradition-bound hill farmers whose "nay" votes on issues stand in his own way. His frankness has been taken for arrogance, and it galls poorer farmers, for it is not in the mold of modesty exhibited by, say, Yankee doctors who come home to practice in shirt-sleeves, Yankee football stars who praise their teams while delivering up game-saving touchdowns, of Yankee lumbermen turned wealthy landholders, who never build mansions and never plague their debtors in public.

With even less perspicacity than his father (both father and son voted for Reagan over Ford in the seventy-six Massachusetts Republican primary), Lee espouses conservative causes.

In the 1976 Conway town meeting Leland voted against assigning some attractive sections of the town's roads to the state's "scenic highways" program because it "limits the landowner's flexibility." He was against denying passage along town roads to nuclear cargos, asking, "Are you going to stop every man with a radium dial on his wristwatch?" And he voted against the successful warrant calling for establishment of a public power corporation. "When you see how bureaucrats in the government handle every other thing they can get their hands on, you can't see handing them the electric too."

Lee is a man very much in the tradition of the "Yankee breed" his father so favors, the ones who "like to be ahead, but prefer not to show it." He seemed to take a certain delight, during the first months he knew me, in presuming a simplicity, insensitivity, a lack of depth and thoughtfulness that are the opposite of his true nature.

That quizzical grin on Lee Totman's broad face is a smile laden with the delights of high irony, of knowing what the audience doesn't think he knows. He is quick-tempered, and one of the first subjects that comes to the mind of any farmer in the county when the topic of Lee Totman is raised is the tale of an encounter with a fellow we'll call Smith the Semen Salesman. Smith is a successful, efficient, if perhaps boisterous and abrasive, member of the county's "agricultural infrastructure," a man who goes from farm to farm inseminating cows in heat.

Lee (with a sort of guild-consciousness quite in keeping with his conservatism) is an out-and-out farmers' co-op booster, and he buys semen from his co-op. Smith the Semen Salesman sells for a private company. Lee had told Smith more than once, "No, not interested." Smith showed up to try again once too often, and, as Lee himself tells it, "I didn't like seeing him drive in. I was on the tractor, hauling manure from the barn to spread on the lower cornpiece. I didn't do anything much, just drove the spreader right past his car, maybe a bit too close, I guess. . . ."

Rumor has it that the spreader flails were turned on, that the spreader raked the side of Smith's station wagon, and that, to quote Smith directly, ". . . and you can quote me, too, farm's been handed to him on a platter and gone downhill since he got his hands on it, too."

There is a bit of controversy in the area — it's an occasional topic in Laundromats and barrooms — over how much of the steady success of the Totman farm is Lee's doing, and how much is based on Ray's continuing supervision and lifetime savings. The controversy gives local venality a fair display — for it seems to be nonsense, testament only to the harshness of Yankee suspiciousness, to the demanding and competitive milieu in which farmers here operate. There's no denying the luck — or was it Grandpa Frederick's foresight — of having a beautiful piece of land to till. And there's no denying that Ray, wanting nothing so much as to have Lee go ahead and farm, has made it as painless as possible for Lee to buy into the operation, while some other farmers in the county, whose families were more given to a moment's frugality and less given to affection and achievement, sent their boys off to "go it alone."

The truth of the matter is that Lee Totman is a spectacularly good farmer, perhaps better, if that is possible, than his father. And since taking over from Ray, Lee has built new silos, a new barn, cleared acres of prime bottomland, expanded the herd, and raised the herd milking average by over five thousand pounds an animal a year. Lee Totman's cows now give more milk — nearly a ton more milk apiece per year than do the cows of the county's next-best farmer, nearly two tons more than the county average, and over three tons more than the national average.

One measure of the success of a dairy farm is a somewhat elusive figure called "rolling herd average." It basically represents an average of yearly production of each cow in the herd. The national herd average is now around 12,000 pounds (about 8.2 pounds of milk make a gallon). The Massachusetts state herd average is nearer 14,000 pounds — reflecting the necessity for New England farmers to do better than farmers elsewhere just to stay in business. I engage in this statistical excursion because it documents the story of the Totmans' incredible farming skill.

In 1924, when the cows of America were only averaging about 4,000 pounds of milk a year, Ray Totman had his herd milking 8,200 pounds. The national average reached that same figure *forty* years later, in 1965. But in those forty years, the Totman herd kept improving. When Lee began working on the farm full time in 1949, the national herd average was 5,272 pounds per year, and Ray had the Totmans' herd average up to about 11,500 pounds.

When Lee took over the farm in 1961, the national average was 7,290 pounds and the Totmans were milking about 14,000 pounds. Ten years ago, the Totmans' herd, now under Lee's guidance, milked 16,029 pounds a year, while the national herd average had climbed to 8,552 pounds. And in 1976, while the national herd average, under tighter business conditions that were quickly ruining the worst farmers, had climbed to about 10,500 pounds, Lee's herd was milking 18,500 pounds. A year later, it was up to 19,500 pounds.

There are two steel engravings, a matched pair like the masks of comedy and tragedy, in Solon Robinson's 1879 agricultural masterpiece, *Practical Hints for Farmers*, and they sprang to mind the very first time I passed the Totman farm. The engravings, on facing pages, bear the capitalized legends FARMER SNUG'S RESIDENCE DURING HIS LIFETIME and THE SAME PLACE UNDER FARMER SLACK'S MANAGEMENT.

Farmer Snug kept his place tidy. An inspector's checklist would read very nicely: Fences — mended. Barn — neat as a pin and bursting with provender, farmer seen through open mow doorway, working hard. House — upright and commodious, with porch straight and true.

Yankees traditionally build porches that will sag after a decade, and tack them on houses built to stand a century. I think it is a custom smiled upon by church fathers, because it insures that the porch will be a barometer of the morale of whatever occupants may be therein. To knowledgeable Yankee passersby, Farmer Snug's straight porch is the equivalent of a gold star in Saint Peter's ledger, pasted up in the column headed "Hard Work and Keeping after Things." New England is a harsh climate not only for crops but for neighbors and porches as well. Any flagging of morale — any passing of days skulking indoors in a state of depression instead of working diligently outdoors in the soothing earnestness with which the industrious look upon themselves, any slackening of righteousness — and down goes the porch.

Farmer Slack's porch — on the way down. Fences — down already. Shutters — on five of eleven windows in view, shaken free of hinges now. Barn — empty, mow door off track. Pigs — out. Trees — the trees have scarcely bothered to leaf out. And on the sagging porch, an adult male who appears to be able-bodied and of working age lounges indolently.

As anyone who has ever lived in an agricultural region can attest, Farmer Slack dwells in every rural town. Farmer Snug is a rarer bird, but he does live in Franklin County, and Lee Totman is his name. I have tried to understand his phenomenal success. His virtuosity is startling in a way that is hard to communicate to nonfarmers. He is more than just a good farmer. He is *right there.*

He doesn't waste moves. He is always set up for the job he needs to do. He plans only as much as his equipment and help permit. He takes shortcuts where they pay and lavishes attention where that pays better. His manure truck is an old unregistered jalopy; his equipment shed is made of old telephone poles and sheet tin. But his milking time is usually about twenty minutes longer than it needs be if he is in a rush, and he spends hours in seeming idleness, just watching his cows carefully. He has, for the multitude of chores which make up a farmer's day, the sort of sense of essential motion that a Japanese calligrapher brings to the drawing of letters. Things are well set up, and usually go right. When Lee's tractor does get stuck in the mud, he laughs.

The right equipment is always near at hand to get it out quickly and easily. Lee is always at work. He is always on the farm; when he needs something he has it delivered. When he does have to go into Greenfield, he goes for as short a time as possible — he says the buildings seem to shake and vibrate as he passes between them. On Mother's Day and on his wife's birthday, he drives to Deerfield — all of six miles — and gets chocolates from the local pharmacy.

At work he is as happy as a hog with a full trough, obstinate, singleminded and intent — intensely concentrating on the chore of running the farm right. In spite of his seeming hardheadedness, he brings to each day's plan an alertness, a flexibility — in solving problems, in reordering priorities in the light of developing events — that combines a realistic comprehension of the cost of doing things with the drive to do what he finds ought to be done. He does this with a stern consistency, with a seemingly carefree, silent and smiling staunchness day after day, decision after decision. He dwells in the world as one who fully accepts personal responsibility for each of his actions; it is the world of the just, which by the way he sleeps the sleep of, snoring some.

Lee stays home and makes things work right because he *can* do it; the perfection of the farm expresses and proves his being; away from the farm things must seem unsatisfactorily wild. As a consequence of his success in building a world that yields to his logic and his effort, he doesn't give much of a damn what other people think and say about him. However, he enjoys appreciation as much as he despises sloth on those few occasions when he allows it to cross his path or inconvenience him. He has a few friends, with whom he and his wife visit once in a while. They are people whom he finds good cause to admire.

Lee is the apotheosis of capitalistic America. If he is admirable for his organization, his concentration, the focus and death-defying outpouring of constructive energy, and if for these qualities he seems to some to be endowed with a Zen-like aura of certainty, of instinctual correctness in every act, there is a wonderful irony to that thought: Lee Totman's organizing precept, central referent, the wellspring of his ways, mother lode, his straight, his narrow, his guardian angel, his pole star, is one single elegant question.

What's the payoff?

He does not seem to be an avaricious man. It is a spiritual question. If the answer looks good, he goes to work. "I'm not so ambitious," he is fond of saying, "that I like to take extra steps. There is too much else to do. Routine is what's important around here. Perhaps I'm good at routine, and if I am it's because I'm lacking in imagination. I see that some of the very intelligent people I know concentrate on many goals — they should never farm because it would restrict them too much. I'm pretty happy just working at this one thing."

Perhaps it is his immunity to innuendo, to criticism of any agricultural sort, his impeccable standing as a first-rate farmer beyond even a suspicion of posturing that so aggrieves some members of the farming community. For Lee, as for Ray, will works. Lee imagines that will works for everybody, which of course it doesn't.

Lee has me home for supper. We walk down the road from the barn to Lee's house, squinting in the brightness of the afternoon sun. It's one of those fall days so clear that it prompts thoughts of fog and winter storms. The house is deserted when we get there. Lee's wife, Betty, is off at a church meeting in Colrain. She travels the twenty miles to a Baptist meetinghouse there several times each week, because she likes the seriousness with which the minister takes his mission. She has left us a refrigerated feast — chicken salad, fruit salad, potato salad, and green salad. Also milk and cupcakes. Lee and I eat, and as we eat, we talk about his family, about cows and about the weather. Lee is not a churchgoer. His wife wishes he were one. Lee's son Gary is away at college in upstate New York. He is on the golf team there, and is not sure he has enough interest in farming to come home and wait his turn. Lately, it seems more likely he'll be back. Gary's sister, Karen, is out with friends this afternoon. "She wants to get a van after high school and drive across the country. My wife wouldn't like to hear me say it — but I can see where that would be one whale of a good time. I never did anything like that. Never left here."

Lee says his cows are doing well on a new computer-balanced feed ration program he has just bought from his co-op. He says the weather has not been too bad this fall, that he takes what comes along, and that there's no point complaining about the weather. Eventually when we are done talking about the family, cows and the weather, we inch up to ideology.

"A fellow was all but forced to milk twice as many cows in the same length of time, once milking machines came along," I say, "just to stay in business. Had to trade in the horses for a tractor. Automatic cleaning and feeding are now a must too, not that they don't make life easier. But as soon as farmers had time and power to get more done, they had to go right ahead and do more just to pay for all the labor-saving devices. And do you know what?"

"What?" Lee dutifully asks. He's grinning even as he eats. "That's it for the good old days. No more farm communities in New England. What little farming we have a few farmers can handle. Everybody else works out, forgets farm ways altogether."

"I'll bet you think that's too bad." Lee's grin broadens. "But to me that's business, free enterprise — all that, isn't it? The ones who go out must think it's better to go do something else — go on welfare maybe, or else the state makes so many regulations they are forced out. And another thing. You think everybody who doesn't farm ends up on welfare or working in a factory? Look at my brother Conrad. He's a college teacher. Writes books. If there were no milking machines or tractors or artificial insemination or automatic gutter scrapers, you know what Conrad would be doing? He'd be in the barn with a shovel."

Lee finds a cupcake, eats the icing first, like a young boy enjoying himself. The discussion goes no further. It may be that alienation is nearly pervasive nowadays, that no one does any whole thing, that almost no one gets the chance to do work he's proud to do, that managers arrange to have things happen, while workers work on fragmented, disconnected chores which may earn them living wages but don't seem to be useful, that don't make sense.

All that may be true, but Lee Totman doesn't have such problems in his world. He does something useful; he turns grass into milk, and he does it wonderfully well. He drinks the milk now as if to emphasize the point. He is the last of unalienated labor. In a national context his attitude is vestigial, an antique even in the universe of farmers. But the fact is clear — even now the system still works for Lee. There is no point in talking to him about my own politics. On the walk back to the farm the sun shines; I breathe in autumn.

II

Surrounding the spectacle of a good farmer working deftly, the sweetness of the harvest air, the winsomeness and pride of an old couple who have done well with their work, their children, each other, surrounding the pleasantness and the political anomaly of a place where the rules still work is, one must not forget, the context. There's no escaping the context. The Totmans stretch the context — they are the tip of the forward tail of the bell curve representing the distribution of good fortune and character within the context. The trends of the history of technology — to displace skilled labor, to force expansion beyond the carrying capacity of the land, to require more capital than one's work can generate, have been turned aside on the Totman farm — it is stronger than ever. The work-weariness of so many small farmers has not visited their farm. The hard New England weather seems to soften to suit them. The ruggedness of the terrain smooths out where they live. The Totmans have turned the forces of history — and of nature — to their advantage.

Even with Lee's shortened daytime milking interval, there are long months in winter when he begins morning milking in the dark, and begins evening milking again in the dark. Lee's suffering is less than that of dairy farmers in the north of Norway, who even eat noon meals in the dark on Saint Lucy's Day. But it does cast a mood of inwardness upon cold season workdays. "If it's dark enough," a neighbor of his once commented, "at least when I look out the barn window, I can't see how poor the weather is."

New England weather is true to its reputation. Winters are always hard, and occasionally, summers improve upon them very little. On hill country farms one can still pick up vague tales about a year a long time back when summer never came at all. It's called "eighteen hundred and froze to death" (according to Harold Wilson's lovely *Hill Country of Northern New England*) and happened in eighteen hundred sixteen. Spring came early. Frost followed on the eighth of June. And snow followed that, in every month of the year. All but the frost-hardy cabbages and small grains failed. River towns turned to fishing, and bartered fish for other sustenance. Entrepreneurs from farther south made killings by shipping in flour. Congregations grew somber and vigorous, savoring, with the proof of half-empty stomachs, the imminence of Armageddon.

The next year things were all right again. Summers seldom have been as bad since, although there have been hard summers. Usually, there is much to endure, but much to enjoy as well.

The amount of damage winter does to grassland depends on the snow. Lee says he likes to see snow covering his land from early in the fall until well into the spring. "The earlier it comes, the better I like it."

Yankees of the old school are an ominous lot, much given to actuarial pronouncements. Their homily suitable to snowless Christmases is: "Green December fills the graveyard." "Take Mr. Fitzherbert," a wrinkled crone once said to me, offering up the late ancient neighbor Fitzherbert as proof of the maxim — "Fitzherbert sat for so very long with nothing to do between corn and the first time he had to shovel snow that he keeled over and died halfway shovelin' to the barn."

Open winters damage fields as well as people. There are two ways of looking at this. First of all, snow does good things. And second, wet and freezing weather does bad things that it can't do if snow is there.

The main thing about snow is that it is a good insulator. In those pleasant years when snow comes before "hard frost," the ground scarcely freezes all winter. Dig down through the crusts and soft layers of snow cover, and on through an inch or two of frozen turf in Mildred Totman's garden, and even in the middle of the quietest, whitest, clearest and coldest snap of January, even when the temperature hasn't risen above minus ten for a week and nights squeak at thirty below, the garden soil will be soft and loose. Other years, snow doesn't come until January, and in spite of the consequences of open winters the graveyards close down.

Snow has shared with lightning the grateful appellation "poor man's fertilizer." There are old Yankee writings that advise plowing in the first snow of the year to feed the soil. Lee finds it useful to leave the snow where it is. Many grasses are frost-hardy, legumes less so, and alfalfa, queen of legumes, is the least frost-hardy of all the common forage plants. Ray says a fresh fall seeding of young alfalfa will stand upheavals — the thawing, refreezing and thawing again of open weather — better than long-established stands. He says the shorter taproots of young plants will ride with the shifting soil, while the taproots of older plants, reaching deep into the earth, will tear. As long as snow comes early and stays, though, alfalfa, young or old, will ride out the winter seas safely.

Lee's neighbor, Graves, who has tapped the Totman maples since Lee gave up sugaring to concentrate on the dairy some years back, is equally glad to see snow come early and stay in place. Sap flows late in winter, as snow melts, and not continually, but in daily "runs" during bright periods of freezing nights and warm days.

There are a hundred theories on what conditions bring on rich sap (windy mornings, full moons, trees along roads, still mornings, waxing moons, trees off by themselves) and the second fifty of these theories contradict the first fifty. Concerning the one-hundred-first theory there is no dispute, and it is this: early snow cover makes for rich sap and a long sugar season, one that starts early and endures late into a cool spring before the sap boils up dark and buddy and it's time to stop. As one of Lee's neighbors, Jim Manwell, explains it, "Roots warm up faster and there's more time for the sap to flow before it goes by, end of the season. Tree doesn't need to eat so much of itself during the winter, either — has more left to put in the sap."

Snow protects bare, newly harvested fields from washing. Under it a carnival of winter animal life lives within the special ecological niche called subnivean — between snow and earth. The living community includes arachnids, mites, and even small mammals, meadow voles, deer mice, and other rodents. They thrive in the protected climate. However, a single pass of a snowmobile, ironing a field a foot above the heads of an active subnivean community in a biologist's experiment at the University of Minnesota at Bemidji, resulted in the utter disappearance of that community (the subniveans, not the Bemidji). Lee Totman is an avid snowmobiler, if a bit more solitary in his pursuit of the hobby than most. He stays off his alfalfa seedings, where the ice under snowmobile trails is fatal.

Instead he rides on official snowmobile club–maintained routes like a good fellow. He spends the best part of the winter working — logging for lumber to use on the farm (most of the farm buildings are homegrown), constructing new buildings, repairing old buildings, watching the cows, feeding the cows, and milking the cows. He is always glad to see spring arrive, if it doesn't come too early.

"If you don't like New England weather," Mark Twain intoned, "just wait a minute." Conway's growing season runs at least 120 days between frosts, from May 15 to about September 15. A brilliant month of fall, and it's winter. There's never much more than a yard of snow at a time on the ground, but in most years once snow comes it stays, and it leaves late. Winters go on and on. Mildred Totman's crocuses rarely bloom before mid-April, by which time they have come and been gone a month in the tropics of New York City, a five-hour drive southward. After the crocuses, summer comes quickly. On either side of the New England growing season sit periods marked by varying weather. And during these times, New England grassland keeps right on growing. Lee Totman will be well along with the harvest of his first cutting of hay before his corn crop — planted to miss the frost — is fairly emerged from the ground. Because of the long grass season, well-tended Yankee hayfields yields up to five tons — about five hundred dollars' worth — to the acre, the season.

What causes the bountiful hay yield is the wet, which comes to fields in two forms, "runoff" and "ground water," the one on, the other under the surface. Yankee farmers read wonderingly of extreme drought in Iowa and California. A classic text of New England field management is called *Grassland Farming in the Humid Northeast.* Farmers have more to fear from summer hail battering a half-grown corn crop than they do from August drought drying out hayland before winter feed is in the barn. (Big Glen Schmidt of Buckland did once lose two acres of fine corn to a tiny twister.) Unfortunately, the same rains that make grass grow better in Massachusetts than on the prairies, that send winter rye up head high by the first of June, also make it hard to dry hay. Fields should be cut in "early bloom," when shoots are still young. But the time of choice for cutting first hay is also the time around Conway when May showers are most likely.

New England farmers try to shorten the time hay must be left to dry before it can be brought into the barn, and they do it in a number of ways. A machine called a tedder is usually run through drying hay a morning after it's been cut. It exposes to the sun grass previously buried under other grass. It has thin fingers that toss grass around. A tedded field of hay will dry a full day earlier than an untedded field. As one travels west in America, summer weather grows drier and the tedder less common. By the middle of Iowa, many farmers have never heard of tedders. Lee seldom mows with the sort of simple sickle-bar mowing machine that old Joshua Totman used, although it adapted well to high-speed tractor work and was the machine of choice into the nineteen sixties. He has switched to a mower-conditioner, or "windrower," which severs stem from root, then passes the stem back through heavy steel and rubber rollers that crimp and abrade it so it wilts in the field more quickly. The mower-conditioner, at some capital expense, saves farmers with tractors powerful enough to run them a day of field drying time. Blessed with days hot, sunny, and windy, and fields neither too thickly planted nor too dank from still weather or underground springs, Lee Totman can mow as soon as dew is off the grass in the morning (his great-grandfather, scythe in hand, would have favored dewed grass), and may be able to bale up nearly dried hay the following afternoon just before evening dew.

During first cutting, one is given good weather only about half the time. The other half, hay gets rained on. Hay rained on just after it is cut, when it still looks green and alive, will come to little harm. But if wet when nearly dried, it soaks in the fresh moistening. It loses quality like spent tea leaves. Ray says it ends up "having less milk" or "less goodness" in it. Protein content decreases. Vitamins dissolve. When the hay finally dries it "shatters," loses leaves as it is raked, and in the leaves lies the goodness. When farming was less business and more way of life, the problem of wet hay seemed less crucial. But nowadays a farmer must be jealous of every mouthful his cows consume. Cows fed poor hay make less milk, and milk checks keep the banker at bay.

On the Totman farm, of course, the banker is so far at bay that he appears as a friend and not as adversary. One reason why is that Lee does not make hay until first cutting has come and gone, until the predictable dry spells of July and August when he can mow a second, a third, and a fourth cutting (that's two cuttings more than most local farmers make) that please him. First cutting goes into the silo as haylage. Haylage needs only to wilt on the field for an afternoon before it is gathered, chopped fine, and blown up into the silo. Haylage can be stored at about 50 percent moisture, baled hay at about 10 percent. Lee has recently added a silo to handle haylage. Unlike hay, haylage can be fed out mechanically, by the same machine that feeds cows corn silage. Haylage ferments and preserves grass with its virgin nutritiousness nearly intact: it is the way around the Yankee weather problem. It takes an investment and skill before a farmer can set about making it. It takes a head on a fancy corn-silage chopper to handle it. It also takes a silo to store it. For automatic feeding appropriate to a one- or two-person farm, that frequently means an upright silo. Lee's new cement stave silo cost about twenty thousand dollars. It will, of course, pay back respectably, or he wouldn't have done it. But on most New England farms, it is not the first place a farmer returning capital to the operation thinks to put his money.

Lee makes haylage when weather is most uncertain, hay when things are predictable. (Although, Ray maintains, "In dry weather, all signs fail.") But even then, Lee leaves less to chance than do most farmers. The shed behind the loafing barn where he stores hay is fitted with a slatted floor. An open space under the floor leads out, through a huge port in the shed wall, to a fan the size of an airplane propeller. If weather threatens, Lee frequently gathers in hay that would otherwise seem damp enough to heat up and set the barn on fire. The fan draws fresh air through the bales.

During the few days of harvest the entire investment in crops pays or fails. Then, more than at any other time of year, time is money. A farmer who spends half his harvesttime working hours loading and unloading a hay truck — as so many

Yankee farmers still do — will bring in less hay between the rains than will his more mechanized neighbors. Lee Totman and his father have devised a hay gathering system worth imitating. To save the "boys" needed to toss forty-pound bales onto a truck and stack them there, Ray bought the first bale thrower in the region. A wagon is towed behind the baler. Hay bales fly from baler to wagon.

Lee uses another innovation as well. He makes his bales half-sized, and adjusts the baler so they are bound loosely — unlike most farmers he wants the bales to distort easily. When the wagon is full up with sloppy twenty-pound bales, he drives it up his road to its very end, high up at the back of the barn, and unloads the small bales rapidly onto a bale elevator, a gangway with a spiked chain running along its center line. Bales impaled along the chain are drawn into the hay shed. On most farms, at this point, hapless laborers haul bales double the size of Lee's off the elevator and stack them neatly in lofty haymow chambers. In Lee's hay shed, bales are off-loaded from wagon to elevator at roof level, travel in just under the peak aboard a horizontal elevator, bump up against a board set diagonally across their path, and tumble down into the mow. Because they are so loose they bend and twist and fill in the empty spaces. The system is elegant, makes use of on-farm technology, is fast and cheap. It saves a couple of laborers, and what may be more important, it saves hours and days that can best be used to harvest rather than handle hay. The savings in hay quality and hay-not-spoiled-by-rain must be even greater than savings in labor costs.

To make hay at all, much less to make it well and to get it in with style, is an act that goes against the laws of nature. The same rain that waters rich grasslands half the year around also supports forest plants. And in New England, field plants lose competitions with forest plants in just a decade. New England wants to be forest. In the earliest days before land was cleared by the Indians and many generations before explorers invaded and Empire settlers colonized the Indians' clearings, the region's only naturally occurring grasses were marsh grasses and salt hay. The earliest white settlements centered around seaside, riverside, and island barn sites. Natural freshwater marshes — extensive ones occurred around Hartford, Springfield and the lower Merrimac River — drew settlers. As Betty Thomson points out in *The Changing Face of New England*, it is flooding that drives the pervasive forest from this rare natural mowing land. Salt marsh floods twice daily, rivers seasonally. Some southern trees have learned to grow on floodland, but in New England, the marshes alone are given naturally to grasses.

The grasses that so trouble Lee Totman by invading stands planted with tenderer species, the grasses we now call "native," are not native, then, at all. Ms. Thomson quotes an early settler named John Josselyn, writing in 1672 to say over forty different weeds "sprung up since the English planted and kept cattle in New England." There is some thought that before these jostling imports arrived, fields — and woods too — supported varieties that did not exhibit the same glaring struggle obvious to us now — that the protocols were mostly settled, that a stable population of triumphant species went about its business less conspicuously at war.

To manage the open land on a farm successfully requires a tactful and energetic exercising of human will over the natural course of wind, rain, and time. What Lee Totman's beautiful hayland wants to be is a maple-beech hardwood forest. Given a respite of just a hundred fallow years, that's what it would be again, arriving at this climax condition after incarnations in transit as fields of moss, strawberries and blackberries, then as stands of quaking aspen, white and yellow birch. Finally, hemlock and white pine, ash, white and black birch and cherry would move in, never quite to leave, but to grow and fall before the aggression of great broad maples and smooth-skinned beeches. Today fusarium wilt attacks the maples and "split bark syndrome" decimates stands of beech. The only near-virgin forest to be seen is an awesome museum piece, the woodlot of the Harvard Forest at Petersham, Massachusetts.

Nature on its own arranges things in other ways than Lee Totman does. Were the farm to vanish, the soil on the forest floor would soon have a pH of about 5.5 (quite acid) and would support a different microbial community from the ones that live in Totman's stately and often-limed fields of alfalfa (pH — a neutral 7 on the nose). The killdeer, robins and cowbirds who like plowed ground would give way to the woodcock, wild turkey, and partridge of the deep woods. Hare and squirrel would be fewer, bear, elk and wolves would resume their residence. This is how the forces of nature had things worked out for tens of millennia. This is northern Massachusetts's self-maintaining condition, maple-beech forest that won the good Lord's competition to stay in place. On the Totman farm, maple-beech is the state known as "the scheme of things."

To undertake to diverge from it is to assume a great undertaking indeed. The modern world, as represented by Lee Totman at work on his tractor, shoulders this chore as a matter of course although every step wades against the inexorable tides of nature. The war against the jungle is manifest off the farm in such simple sights as lawnmowers and cement sidewalks. We humans wage this battle with a blithe finesse, with a heedless and unself-critical expenditure of energy that might signal to some wise observers in flying saucers that *Homo sapiens* is a race of very lively creatures, given with pluck and abandon to great works that make things different from what they would be otherwise.

While Totman's dairy farm does not exactly dare the encroaching jungle, as might some aboriginal rain-forest clearing, the spirit of the affront is the same. The main difference is in latitude. In the north, nature moves with softer, slower, and gentler but no less persuasive force than it does in the tropics. Thawing and freezing; supplanting of species by tougher species that slowly edge in, causing stress, disease and slow death; grinding down by course of wind and water and snow— all replace the fierce tropical battles of rain, heat, stingers, poisons, thorns and stickers, horny hides, raging floods, armies of ants and parasites that eat neighbors in minutes, which together spend the additional energy of the tropical sun. But whatever the intention, whatever the latitude, the battles of nature are the same; the methods up north are somehow more in the spirit of slow, considered and vindictive Yankeedom. The land shapes the people and the people shape the land.

The Totman farm now consists of about a hundred acres of cropland and pasture, and about two hundred acres of woodlot. After Lee took over, he cleared about ten or fifteen additional acres of bottomland. It had grown back before the lower farm was annexed to the Totman farm in the 1940s.

A hundred years earlier, between 1820 and 1840, virtually the entire hill where the farm sits was cleared land. It happened during a brief flowering of Yankee agriculture that now appears rather magical. The boom started in the 1810s, when roads broadened the region's commercial life, and it grew by bounds in the 1830s and early forties, when railroads and mills spread along the valleys and riversides.

The crop of choice turned out to be sheep, especially the merino sheep first imported in large numbers, along with the native shepherd from the Spanish Escorial royal flock, by one William Jarvis, American consul and chargé d'affaires in Lisbon. Farmers whose villages had been sufficient unto themselves turned for the first time to making money. By 1830, half the farms in Massachusetts were said to be mortgaged. Boston's supply of fowl, butter, and eggs nearly ceased in 1837 because most farmers had turned to sheep. Fortunes were made selling wool, breeding stock and meat, and other fortunes were made at the mills. In 1842 President Tyler, pressured by mill owners, altered a tariff law, allowing the importation of European wool. It sold at prices farmers at home could not match.

The boom faded, sometimes leading farmers into tasteful retrenchment in other agricultural pursuits, but more often leading them into poverty as wool prices failed to meet farm costs. Many farm families joined the migration to lands not yet depleted by intensive agriculture, lands that lay over the Alleghenies to the west. As the Western Reserve was settled, the cleared woodlots of New England began to close in — maple-beech on the rise. By the time Joshua Totman moved into Conway in the 1860s, the most marginal rank of farms was already abandoned. The mill owners continued to prosper working the imported wool; many farmers' children moved off the land and took mill jobs.

The oldest folks in Conway remember when a few more of the now wooded hills were cleared land, but the time when most of the land in the area was clear has passed beyond memory. Fred Call of Colrain, just north of Conway, tells of his father complaining about the shame of abandoning productive land in the mid-1800s. Today about 70 percent of the state is wooded and the situation is quite stable. Land lost to farming nowadays — 10,000 to 20,000 acres a year of the ground that's still farmed — is lost chiefly to roads, housing developments, and shopping centers that sell food imported from other regions. Massachusetts imports all but 15 percent of its food. Even milk is shipped into Massachusetts. The only crop the commonwealth produces in excess of local wants is cranberries for Thanksgiving. It happened because neglecting the land paid better than caring for it.

A pasture constantly used and seldom replenished — there are many like this in New England — will eventually become a hostile place for plants nourishing to cattle. It will become a hostile place, in fact, for plants even accessible to cattle. Rundown pastureland constitutes a unique chance for some plants to thrive that wouldn't otherwise be in the running at all. Open-field grasses, especially volunteers of seeded stands, have been bred to compete best in very fertile conditions. In improverished soils, they start and fail, giving way to varieties that have a penchant for wresting a living in harsher circumstances. Sour land is a particularly poor environment for most legumes. Were they able to seed themselves, they would resupply the soil with nitrogen.

But the condition that is perhaps hardest of all on pasture flora is the presence of cows. Cows modify the population of plants that grows in their fenced-in dominion as surely as does infertility or acidity of the soil. Cows eat the most succulent plants first. If their pasture is large and they can afford to be choosy, untoothsome species will ripen and then lose their feed value uneaten, but will go to seed.

Some plants prosper in pastures with cows that would fail if the cows were removed. Their advantage over other plants is that cows don't care to graze on them. Mullein leaves are covered with a velvety skin. It is hard to anthropomorphize the tastes of animals given to feasting on dry hay, but cows don't seem to like these hairy little leaves. Several other varieties — ones that might taste fine to cows — are armored against bovine aggression. These are the sharp plants, especially hawthorn, juniper and thistle.

Juniper aggressively reroots itself from its low branches as it expands, has spiky leaf tips, which ward off hungry cows. It blocks light from plant competitors that would dominate it in less acid ground. It can cover acres at a run. Hawthorn requires more fertility, and cows help out. It starts up in the fertile comfort of a cow dropping, and drives away grazers during its ascension by means of spikes as long as toothpicks and as sharp as glass splinters.

These species must be to cows what porcupines are to country dogs — tempting, but unavailable. And cows act as dutiful groundskeepers, mowing away palatable contenders while leaving the forbidden fruits to prosper in the adversity that is their proper garden.

Pastured animals physically damage the fields they graze. They nip in the bud the sweetest grasses. Cows' hooves trample and crush young shoots. Fresh manure first burns and then mulches the ground it later nurtures. Good pasture is harder to maintain and harder to use to advantage than good mowing. A pasture offering prime feed one week may have overripened the next. Keeping up a proper succession of pasture grasses was always an art; Ray Totman was particularly good at it. Hay can be taken when it is at its best, a field at a time. Occasionally an unusual farmer solves the pasture succession problem by fencing pastureland into many small lots, and moving the herd from lot to lot almost daily, leaving grazed land weeks to recover before it is used again.

Lee Totman has far less pastureland than Ray did; the areas he does keep in pasture are the ragged, ledgy lots at the sides of his squared fields. He uses these to store animals that are not making milk but are simply waiting to get older or waiting to give birth. Lee does manage to lime, fertilize, reseed, and even on occasion to clip down the bits of gone-by and cowproof herbage that threaten the quality of his pasture seedings. He does it when he is fertilizing and liming the cornland and mowings. "It saves me more than it costs, in the long ran, I guess," he says.

If it paid Ray, earlier in his farming career, to pasture animals, and to put long working hours into improving the feed growing in the pastures, the same practices would break Lee, were he to have retained them. The increased cost of building a fencerow, the increased efficiency of grass harvesting equipment, the improvement in the nature of barn design, allowing cows to feed themselves indoors as well as they once did grazing, the competitive milk marketplace that profits only the most careful farmers, all demand that cow-feed be grown in fertile fields, then brought to confined aimals. Under these conditions, the challenge most dairy farmers in New England find most difficult to meet over the long haul is the challenge of maintaining soil fertility. Lee's program to maintain his soil begins each spring as soon as the snow melts from his fields.

On the quarter-mile walk from house to barn each dawn in the weeks following meltoff, Lee veers into the upper corner of the upper cornpiece and scuffs at soil with his boot. The first week the soil is hard. The second week the earth begins to thaw, growing steadily stickier and wet underfoot. Days run bright and warm, nights windy and cold. Sap flows into the buckets along the road.

The alternating freezing and thawing of the ground each spring gives New England soils a wonderful natural "tilth." In fact, to save harrowing seed into a mowing that has suffered winterkill, and to get the seed into the ground before it is shaded by returning grass, some old-fashioned farmers broadcast a "frost seeding" on top of the year's presumed final snow. They count on erratic spring weather first to bury and then to germinate the planting.

After the ritual of prodding the upper field with his foot each morning has gone on for some weeks, with the toe digging into progressively drier soil, the day finally arrives (it was the tenth of April last year, and farmers a few miles downhill in the Connecticut River valley had already been at fieldwork for weeks) when Lee likes what he kicks well enough to drive a tractor across it. Several major spring tasks get under way.

The mowings — delicate alfalfa, stately timothy, intrusive witchgrass alike — have survived and turned, in the course of a few days, from dead brown to lively green. The plowed ground has sprouted a crop of boulders, driven up from a seemingly perpetual store by spring thawing and freezing. Lee works for a few afternoons with tractor and stoneboat, removing the stones. A famous Yankee story fits the occasion, about the city fellow who stopped his car to chat with a farmer one spring morning.

"What are you doing, farmer?" asks the city fellow.

"Picking rocks," says the farmer.

"How did all those rocks get here, in the middle of the fields?" city fellow wonders.

"Glacier brought them here and left them," the farmer explains.

"Glacier? Where's the glacier?" asks the city fellow.

"Gone back after another load," says the farmer.

With rock picking out of the way, Lee turns to spreading manure. One of the few advantages of open winters is that farmers can spread manure for longer than they usually can before snow interferes. Lee has piled a great mound of manure in a storage area below the heifer barn — the heaps look like a scale model of muddy Berkshire hills — and he and his son Gary, home from college for Easter holiday, set to work. They hoist the piled manure with a front-end loader and slosh it down into a huge tank-bodied manure spreader. Then they tow the spreader across field after field, painting wide brown swaths of fertility on plowed ground and grassland alike.

> Manure was once an asset to a farmer. Now it's a liability. Though manure is beneficial as a plant food, it is frequently more expensive than other sources of fertilizer, and disposing of wastes without contributing to pollution has become a significant problem on livestock farms.

So says the tenth edition (1973, borderline of the energy crisis) of *Doane's Farm Management Guide,* the august reference that Wall Streeters consult for the latest on agribusiness management practices.

Organic Gardening and Farming Magazine disagrees: it calls manure "organic gold," "Nature's own fertilizer," and other worshipful things, all of which generally imply that there is a scatological road to world salvation, and that the Way is through composting and lavish application of manure.

There is support for such a position. For thousands of years, manuring has kept Europe's ancient fields from becoming barren. Edward Lisle, the English gentleman who wrote *Observations in Husbandry* in 1757, proclaims,

> . . . where grasses are, that naturally grow in barren grounds, such lands want manuring, and then the better sort of grasses, which carry strong roots, will easily overcome such poor grasses. . . .

In America, there have always been reservations about manuring. The reservations, like those in *Doane's Guide*, are never about whether manuring *works*, but about the cost of manuring versus the cost of taking other measures.

Manuring is one response to a universal agricultural problem. After virgin ground has been farmed for six or eight years, crop yields begin to diminish. There are actually several choices about what to do next, and the correct one is not always obvious. One traditional response is to stop using the ground — either by stopping farming or by farming elsewhere. Another is to mine the land, to shape agricultural operations so that slackening yields are acceptable. The third is to fertilize, and it is the choice most commonly practiced these days, although that wasn't always the case.

Even if a farmer chooses to fertilize, it is not self-evident which of alternative means of treating the soil is best. Farmers throughout history have used all of these solutions at different periods, depending not on whim, belief, or science to determine their choices, but on cost and practicality. Shifts in farming practices mirror shifting economic conditions, and in spite of many agricultural historians' unself-critical notions that those slow to adopt new innovations are hicks and half-wits, it seems more sensible to assume that most farmers have always done what made sense, and then to set about finding out why.

The Reverend John Blake, writing from New York a century after Lisle, is one of America's many dogged manure adherents. In his friendly *Farmer's Everyday Book*, he poses the rhetorical question "How Can a Farmer Become Rich?" and in a rather imploring tone answers it, "By cultivating less land and manuring it more highly, so that an acre will require less labor than heretofore bestowed upon it, and produce double. Make it rich and mellow. . . ." It is a capital-intensive answer for an age not yet completely geared to such solutions.

Even in the Reverend Mr. Blake's times, many farmers owned more land than they cultivated at any one time. Frequently the least-cost solution was to let unproductive land go for grazing, and clear new fields. Other farmers, tens and hundreds of thousands of them, in fact, found it cheaper, in the wake of the sheep bust of the 1840s, to abandon Yankee farms altogether as the land ran down. Even in the mid-nineteenth century, in America, it was not uncommon to find barns for livestock built astraddle creeks and ravines, and built there precisely so that manure could be disposed of at low cost. These barn builders shared *Doane's* view that manure is a "liability."

By the end of the last century, farmers were by and large using their manure to improve soil. Following a craze for South American guano, the commercial production of chemical fertilizers began and, as before, manure again became a liability. The 1906 Yearbook of the U.S. Department of Agriculture says,

> . . . atmospheric nitrogen can be oxidized under the influence of electricity. . . . These direct processes of securing nitrogen will be rapidly improved, and what has been accomplished already in this direction should remove the last vestige of doubt that we shall be able to secure at a reasonable cost all of the immediately available nitrogen we may need, in addition to the great supply that may be secured through bacterial action.

It was the beginning of the energy crisis.

Even the Massachusetts Indians, blessed people of the earth though they were, did not handle the problem of decreasing soil fertility by manuring — not even with fish in cornhills — in spite of the myth to the contrary. They may have bred corn in sophisticated ways, and cropped fields extensively and well for harvests of beans, squash, pumpkin, jerusalem artichokes, and berries as well as corn. But when it came to the problem of what to do when land got poorer, they, like the "bad" farmers of early America, moved on.

Their actual response to the soil depletion problem is interesting in the light of the fish fertilization myth. It turns out the Indians would farm a field for six or eight years, then fallow it for as long as forty or fifty years, then farm it again. It's a measure of tribal stability that they could do so.

They never fertilized with fish. They ate fish. They salted fish for winter. They made needles and fishhooks from fish bones, ornaments from fish scales, and were even apt to draw pictures of fish on rocks. But the Indians of Massachusetts handled the soil fertility problem with a more modest strategy.

A clearsighted anthropologist, Lynn Ceci, when still a graduate student at City University of New York, published an astonishing article a few years ago in *Science*, called "Fish Fertilizer: A Native North American Practice?" It seems that one "Squanto" (whom we have all heard about in school) strode into Plymouth on March 16, 1621, and greeted the Puritans with the unlikely salutation, "Welcome, Englishmen." In English. And right after the first hungry winter ashore. One diarist says Squanto "directed them how to set corn, where to take fish, and to procure other commodities." They did as he suggested.

The following winter, a letter that still survives was mailed home from Plymouth. "We set last spring some twenty acres of Indian corn and sowed some six acres of barley and pease," the letter says, "and according to the manner of the Indians, we manured our ground with herrings, or rather shad, which we have in great abundance." The letter, according to Ceci's account, comments on the "manner of the Indians," at a time when no colonist had set eyes on an Indian corn planting operation. Plague among Indians had canceled 1621 tillage in the region, and when they did plant in 1622, they didn't use fish, much to the Pilgrims' surprise. They needed their fish catch to eat. They had no metal tools. They had no wheels, and therefore no carts or barrows.

Nevertheless, it was accepted thenceforth that Indians fertilize with fish. By 1916, Ceci says, an anthropologist named Wissler felt able to state authoritatively that fish fertilizer belonged to an "aboriginal maize culture complex everywhere in the Mississippi valley and eastward." In 1939, the anthropologist R. Flannery was the first to complain that there were no references to fish fertilization in the ample colonial records of Indian tillage practices. In 1957, a geographer named E. Rostlund seconded the notion, arguing that fish species weren't always available and that the settlers' references were vague.

Yet the fact remains that the colonists did start in fertilizing with fish (which worked well), and that they did learn how to do it from Squanto. Where did Squanto learn it? From Europeans. Squanto had spent his adult life as a kidnappee, and, eventually, companion to explorers and fishing parties. He had spent many years, before the settlement of Plymouth Bay, abroad. In 1614 a Captain Thomas Hunt captured Squanto and sold him for a slave in Málaga, Spain. He may have been kidnapped previously, in 1605 by a Captain Weymoth. From Spain, he was bought and brought to London, where he spent two years on a farm called "Cornhill," owned by an executive of the Newfoundland Company. The gentleman then hauled Squanto over to the Cupids and the Guy Colony at Placentia Bay, Newfoundland. Squanto's luck changed. He was later a pilot and guide for another English captain, who dropped him off on Cape Cod. He wandered into Plymouth, met the settlers, suggested fish fertilizer to them, then died nearby in 1622.

Ceci goes further. He quotes contemporary sources saying that the Indians did not use fish. "Of manuring . . . they know nothing." And, "they are not industrious, neither have art, science, skill or faculty to use either the land or the commodities of it, but all spoils, rots and is marred for want of manuring."

On the other hand, it seems there are records of Europeans engaging in fish fertilization since Roman times, and there were even published accounts of the practice in British journals before the colonists' departure for the New World. The English manuring tradition was extensive by the time Squanto hit town, and he was around at least three manured gardens — some manured with fish — before he returned to America. Ceci concludes, "Squanto acquired his agricultural knowledge from European examples. Then on his 1621 visit to Plymouth, he merely passed along practical advice he knew to be successful from his most recent experience with Europeans, not Indians."

Like the Plymouth colonists, and in spite of Doane's dismal observations, Lee Totman does manure his land. If manure is a "liability" to him, it is one he turns to the best advantage he can. Manure is good stuff — not the cheapest or the easiest- to-handle good stuff available, but good nevertheless. The fact that it is delivered to him every day forces his hand. It is good for the land. Manure is composed of fibrous material, carcasses of millions of the microfauna that eat cellulose in cows' stomachs, of water, and of chemicals and vitamins good for plants. Its application, unlike the application of large doses of commercial fertilizer, improves the friability and absorbency of soil, and releases slowly and in modest doses chemicals that in larger or more volatile applications alter the soil's microbiological community. The problem is that it is sloppy, and for the amount of "goodness" in it, it is heavy.

Each of Lee's cows produces about a hundred ten pounds, about a dozen gallons of manure and urine each day. In Lee's barn, bedded with sawdust like a butchershop floor, this means an accumulation to be scraped up and hauled off each winter day of about three tons. There is plant food in manure — the basic three, N, K, and P of commercial fertilizer. But there is surprisingly little of this in relation to the high costs of gathering, storing, and spreading manure, considering its "golden" reputation. In Lee's daily three tons of scrapings are about thirty-three pounds of nitrogen, six pounds of phosphorus, and thirty pounds of potassium. This is the same amount of nitrogen and potassium as and six times less phosphorus than would be found in two sixty-pound sacks of 20-20-20 from the local feedstore. At the moment it is far cheaper to spread fertilizer than manure, for all the harm fertilizer may do to the soil and all the good manure may do. Lee spreads large amounts of both.

Fertilizer manufacture is energy-intensive, however, and the time will probably come when crop rotations and interplantings, including nitrogen-fixing plants, composting of municipal wastes, and careful use of manure, will be cheaper than the current solutions to the problem of decreasing soil fertility.

Manure spreading completes a cycle that links cow to field, returning the the land what has been grown and consumed from it. The cycle is quite closed when cows are pastured. Animals on pasture, taking all their sustenance from the land, then eliminating on the premises, are said to return about 80 percent of what they remove. The rest is exhaled, or leaves the field as hair, bones, eyes, hooves and roasts. Milking animals return less than beef cattle, the difference going into the milk pail.

When Lee Totman scrapes the barn free of cow manure each morning and tows the spreader from barn to field to complete nature's cycle, he is housekeeping for a modern kind of beast whose very existence seems a special dispensation on behalf of humans. Although no one yet has succeeded in housebreaking the cow, it is nevertheless an emblem of human ambition to alter nature. Cows have been domesticated for millennia — there are ox remains in Neolithic Swiss lake dwellings. The result, as distant from the wild beast called

Bōs taurus typicus prīmigenius as the poodle is from the coyote, is a reflection of nature at her most genial and compliant.

The essence of the cow, if indeed at this point wildness and not domesticity can be considered its essence, is its fugitive nature. The shyness of the mammal group *Pecora* is shared by sheep, goats, antelope, deer, and giraffes (whose tongues are about a foot-and-a-half long). Like these other animals with hooves, cows want to flee when threatened, and use their horns only as a last resort. This propensity for flight is obvious in deer, antelope and gazelle, which all run fast, says the gloriously correct *Encyclopaedia Britannica,* and in mountain sheep and goats, which climb with great alacrity and very slight provocation. Cows on range grow skittish. "Not only do their skeleton and muscular systems form together a perfectly constructed running mechanism, but their digestive system is also elaborately planned so they may hastily snatch a meal. . . ."

All this is less than clear in the case of the modern Holstein in Lee Totman's barn, which looks as awkward hustling down from pasture with a cowdog nipping at her heels as portly women running to catch a train. With cows, evidence of fugitive nature is found in skeletal conformation, rather than in a present-day elusiveness. As the *Encyclopaedia's* own elusive "H. S. P." indicates, the lower leg and foot are long in proportion to the whole limb, lending a stride both lengthy and rapid to the beast. "The surfaces of the joints are grooved and keeled like pulley wheels, permitting free motion forward and backward but limiting the motion in all other directions; joints of this type are very strong and are admirably adapted for swift locomotion over a smooth surface. . . ."

What is essential to the genetic perfection of fugitive animals is that they are sought after, that they live in a world also inhabited by fast-running predators. Were this not so, of course, they would not have been pressured to develop the skill to elude. Every creature eludes predators and parasites in some way; it seems odd that cows' elusive art is running. In an off-hour at least one biologist has suggested that the strong muscular development necessary for good running might happen to be what tastes good to humans; evolution seems in general to define what is nourishing as what is palatable.

The hardiness of such elusive species depends on the enthusiasm of predators for their prey. Predators in search of nourishment are the major selective force determining the genetic composition of fugitives. If it weren't for lions, if it weren't for hyenas, woe on antelope, woe on wildebeests. And if it weren't for people, woe on cows.

In all of natural history, no predators have transformed animals more ambitiously to suit their own omnivorous needs than humans have shaped cows. Probably the process started about eight thousand years ago, probably in Asia (probably its western edge). Probably it spread to Africa and the Middle East in the next thousand years or so, and thence up into Europe. Cows were kept by people who did not yet keep horses. The most ancient fossils of cow predecessors are around fifty million years old, and indicate an animal the size of a dog. A more recent upstart, the aurochs, from which domestic cattle derive, was, on the other hand, a largish sort, weighing in at half a ton.

Vermonter Dirk van Loon (whose fine sourcebook *The Family Cow* treats some of these facts and speculations) offers an appealing look into the nature of the ancient moment when ties first bound cows and people. "Eventually," he writes, "people whose ancestors followed migrating herds of wild cattle found themselves in the lead."

The predators didn't stop there, either. Cows have been vigorously bred, their genetic capabilities channeled for the good of people, not cows, ever since that first halter was slipped across that first obliging bovine's snout. The modifications humans have caused are extensive, have altered the very competence of the beast, leaving here on earth an order of animal completely dependent for survival in its present form on the survival of people. If Lee Totman's herd were turned out and left on its own, free of any human interference, few cows would endure the first winter.

What we have bred into the innocent *Bōs typicus* is a capacity to produce milk far in excess of the needs of its own young, a flesh both tender and built of grass (which is cheap to grow and can grow where human food can't), a powerful body, and a submissive temperament. Cows have been shaped by people to be an obliging source of hauling power even while they give milk and fatten.

Breeds isolated from one another were consciously developed in Europe as long as two thousand years ago, primarily as beasts of hurden, and only in recent history as animals for milking. Holstein-Friesians, today by far the most popular dairy breed in America, have been in continuous residence in the Low Countries since before the birth of Christ. Guernseys and Jerseys ripened slowly in the seculsion of the two Channel Islands of the same names — although herd books and breed-wide improvement plans came into force on the islands only in the early eighteen hundreds. Brown Swiss, the oldest dairy breed, are the ones found in the Swiss ruins. The remains are dated at about 2000 B.C.

For all their extensive breeding, dairy cows still bear signs of the wild beast deep in their natures. The fugitive disposition comes out at calving time. Cows prefer to calve alone and, given the opportunity, will bolt sturdy fences to hide themselves in timberlot pasture.

Cows have been bred for docility, a trait of greatest value during their long tenure as draft animals. It is a trait that humans have stressed, but it is deep in the beast's basic nature — in its inclination to avoid danger. In time of stress, cows head toward each other and away from whatever disturbs them, which is the reason a cattle drive works, the reason cattle mill in a feedyard where fences prevent them from fleeing, and the reason that excitement can start a stampede.

Dairy farming itself goes back nearly to the beginning of colonial New England history. Howard Russell, in his recent book *The Long Deep Furrow*, a rambling but thorough history of New England agriculture, mentions a dairy in Rhode Island where a large herd was milked by female slaves before 1700. But this was atypical. It would be accurate to think that most New Englanders, rural and urban, until the end of the Civil War drank milk from a nearby cow, and ate cheese and butter made on farms. Itinerant wholesalers toured farms to buy surplus cheese production.

Starting at the end of the Civil War, and continuing for the next forty years, however, a series of commercial and technological changes occurred that transformed the nature of dairy farming — that made it a particular trade, backed up by a network of supply businesses, and integrated with processors and marketers dealing in remaking and selling what dairy farmers produced. Forty years seems a short time to transform what had stayed the same for centuries if not millennia. But the changes were in fact a rather late effect of the transformations that for several centuries following the inception of the industrial revolution had altered most other areas of peoples' lives as well.

The technological developments that changed farming conditions were closely integrated with developments in commercial facilities. For example, farmers began "breeding up" to improve genetic potential for milk production on a scale never before seen. This "technological" effort was accompanied by a growth of farms that specialized in supplying good genetic material, and by the improvement of roads, containers, coolers, sellers, credit suppliers, all needed to induce higher production. Innovations took place on the farm — in tillage, animal housing, harvesting, crop storage, and feeding methods. Innovations took place in the processing factories, in development of an increasingly productive alchemy to turn milk into cheese and butter.

The development of a commercial market for fresh milk came more slowly than the development of marketplaces for more shippable, less perishable commodities, such as grains, vegetables, and even eggs. With the spreading of railroads and further improvement of roads, even farmers far from growing urban populations began to think of fresh milk as a potential cash crop. The first railroads were built precisely with an agrarian transformation in mind — to turn farmers into suppliers of sustenance for workers near central sources of power and finance. By 1842 a milk train ran into Boston, according to John Schlebecker's *History of American Dairying*. A burgeoning of milk cooling, storage, and transferring technology increased the efficiency of shipping milk.

Once established, what remained for the perfection of the marketplace was increased reliability of the product. The second half of the nineteenth century saw extensive development first of cheese technology, then of railside "creameries" that made butter. The new trade promoted an entrepreneurship not so much of farmers, whose shift from subsistence agriculture for many decades manifested itself in the adding of a cow or two to stand beside the family beast in the barn, but of collectors and processors. At first, the wealthiest of farmers tended to be the people who started cooperative creameries and cheese factories. These institutions soon attracted outside capital, and developed adversary relationships with their suppliers, relationships that have not altered to this day. Two key inventions led the way to the "rationalizing" of dairy manufacturing.

The first of these, the centrifugal cream separator, supplied farmers with a means of dividing cream from milk in the barn, so they could ship just the cream for processing. The Totman family used to haul cans of cream to a local creamery, which in turn would ship butter on the railway to Greenfield on the same spur line that had brought Frederick's first barn out from town for reconstruction at the farm. The cream separator, which made a fortune for a Swede named DeLaval, for the first time allowed for speedy continuous separation, putting an end to batch separation in pans.

Another invention, seemingly innocuous to the uninitiated, protected farmers selling whole milk from the guile of processors. In 1890 Stephen Moulton Babcock perfected a simplified procedure for testing the amount of butterfat in milk. He discovered a sulfuric acid solution that would act as an agent to dissolve casein in whole milk and liberate the fat. One chagrined factory owner commented that "the Babcock Test can beat the Bible in making a man honest." (The quotation is from Eric Lampard's definitive *Rise of the Dairy Industry in Wisconsin* — fine reading for those of a lactohistorical turn of mind.) Babcock, who might have made millions vending the rig, instead freely published the results of his years of work: "In the hope that it may benefit some who are striving to improve their stock and enable creameries to avoid the evils of the present system, the test is given to the public." If there's no statue of Babcock anywhere, the spectacle of a technological innovator in the thick of the Gilded Age giving away his genius certainly merits one. The test he devised for the first time gave farmers an accurate measure of the quality of individual cows, which vary greatly in the percentage of cream in their milk. It sharpened their breeding efforts.

The buttermakers kept honest by Babcock's invention felt pressure, from the very inception of their commercial existence, from another quarter as well. They were undersold — even in the good old days — by the manufacturers of margarine. A vociferous French fellow named Hippolyte Mège-Mouriez, who had earlier worked on perpetual motion machines, and who had, for Napoleon III (a ruler alert to the possibilities of the impending new technology), found a process that stretched wheat to make 14 percent more bread for the hungry masses, in 1869 fabricated a butter-like spread out of meat tallow. Hippolyte called it *beurre économique*, and while it didn't crowd butter out of the grocer's tub, there was enough of a demand for it so that an American factory, the Oleo Margarine-Facturing Co., was established in 1874, only thirteen years after Alanson Slaughter, of Wallkill, New York, had started the first American butter factory. Margarine producers were market-wise long before their dairy competitors — a pattern self-defeating dairymen still perpetuate by allocating only the most parsimonious percentage of their milk checks for advertising. Margarine manufacturers early learned to package their goods in sanitary and eye-catching regalia, and guaranteed satisfaction at a time when butter came from concentrators, slapdash creameries, and uncertain shipping agents, and was dispensed under unsanitary conditions from tub lot directly into the customers' containers. Having good margarine on hand depressed the price of butter and, eventually, improved its quality.

As hosts of new inventions increased the efficiency of processing technology, farmers were impelled to keep up with the times or leave farming to those more willing to do so. This was the period during which Joshua Totman and his neighbors gave up mowing as a nearly self-sufficient team for mowing separately and more rapidly with the use of newly perfected horse-powered farm tools. Steam-driven equipment was seen occasionally in the Conway area before the end of the nineteenth century, and machinery powered by internal combustion arrived with the new century. Rail lines expanded, and local creameries took virtually all the milk that farmers could produce. The period just before the First World War is still remembered as the golden age of New England dairying.

Farmers milking ten or fifteen cows by hand lived laborious but decent lives. New England terrain accommodated many farms of that size, which helps explain one of the most scenic of New England's attractions today. It takes about twenty acres of ground in New England to feed ten cows and their young. Driving through the area, one still sees farmhouse after farmhouse, each surrounded by ten cows' worth of land. When later developments, especially the coming of tractors and milking machines, forced farmers to milk thirty-cow herds, farms consolidated, extra houses were parceled off, and farmers took to tilling leased or borrowed land — a pattern the few survivors are still bound to follow.

What New England dairy farmers grew, and still grow, on that land is hay and corn. A progression of technology, first for loose hay, then baled hay, finally for haylage, made grass farming here more efficient. Corn growing progressed along similar lines. Silos appeared on the farms of a few adventurous Yankee experimenters by about 1870 — some twenty years after European innovators worked out the process of anaerobic forage crop fermentation. By 1880 silos were extraordinary but not rare. By the turn of the century most dairies had them. Ensiling eased the problem of getting quality feed from New England land in the face of uncertain weather. Before the introduction of silage-making methods — one of the early vigorous educational efforts of Extension officials — harvesting was so costly of labor that corn cultivation for dairy use was limited. After the coming of silos farmers could keep more cattle. Also they could think about milking in the wintertime, as quality feed was now available then.

The same railroads that provided marketplaces for dairy goods also brought competition. As milk production improved, the newly arriving genetically superior animals demanded high-protein grain. It was obtained most practically by rail from cornfields thousands of miles to the west. In the first three decades of this century, corn harvested for grain in Vermont decreased by a factor of eight as farmers switched to silage and began to buy grain. This new habit created a one-sided and mortal dependence on western suppliers, oligopolistic wholesalers and shippers who supply them at premium prices. It represented the loss of local control of a crucial aspect of the dairy operation.

Yankee dairying had entered the modern commercial world. The advent of farming for profit instead of survival, destined as it was in the long run to lift many families out of lifetimes of drudgery and to cause many more yet to leave farming altogether, also ended neighborly interdependence, and set farmer against farmer in competition for each other's choice land. It also changed attitudes. For the first time, an entire class of farmers went to their barns each morning asking themselves, "How can I make more money out of this operation?"

The way most obvious to most farmers was, and is, to sell more milk. Improving equipment allows each farmer to tend a constantly increasing size of herd, up to the limit of available land. That limit is so severe in New England that farmers early turned their attention to coaxing more milk from each cow. The long-range consequences of this trend have not affected the region as early innovative dreamers might have supposed, but such thoughts were far from the minds of nineteenth-century dairy farmers.

A controversy sprang up, first north of New York City, where early development of rail lines and cheese factories to process milk from the area's rolling farmland contributed to the growth of a strong dairy farming region. For decades, these farmers had been supplied with young stock by itinerant traders called drovers. The drovers, who indeed drove herds of cattle from town to town, persisted into the present century. Old men still alive remember supplying their herds each spring from the motley collections of drovers.

In their heyday, drovers moved crowds of cattle from the west to the east, and from Canada down into New England — a route that served to introduce fresh blood from the English and Scots breeds encountered along the trail into the American gene pool. Farmers kept their culls to sell to drovers; from drovers they bought smart-looking replacement animals — animals that very likely were their neighbors' culls. By the

1860s a very few farmers had "closed their herds," and were using their own bulls to "breed up," to improve their herds' productivity.

The testy and opinionated John J. Thomas, associate editor of *Cultivator and Country Gentleman*, wrote from Albany, as late as 1868, disdaining this practice in *Rural Affairs*. His advice is contrary on every single point to modern beliefs:

> First of all, the milk dairy men should abjure allegiance to any one strain of blood; it will never do for him to swear by the Herd-book, or to have any hobby of race. Here and there, a Short-Horn (at great price) proves a great milker; and there are individual Ayrshires who do wonders in the filling of a pail; the Alderneys, I think, never. Grade animals of good milking points will be serviceable ones for him; and if he keep his eye open, as every shrewd farmer should, he will find here and there some raw-boned, mis-shapen native animal, who will yield golden returns. Those animals that will give the most milk under generous feeding, without respect to name or lineage, are the best animals for him. Therefore, in nine cases out of ten, the best milk herd is very motley in form and color. . . .

Having warmed up on pedigrees, Thomas turned his disdain on the new-style farmers he had seen who had begun to raise young stock bred on the premises:

> No milkman can, I think, raise his own stock to a profit. Cows can be grown more cheaply in the back country, where milk is at five to six cents the quart, than he can grow them for himself. He needs to devote all his care and his food material to the immediate object of his calling. Young animals are a waste and a nuisance to him. . . .
> The age at which to buy should not range below five, or above eight, and at thirteen or fourteen, ordinarily, it is time the dairy-man should find another owner — whether among

those good people who love to pet old cows, or those bad
butchers who peddle bad beef. . . .

Most farmers of the time, with rare exceptions, followed
Thomas's advice, frequently for no other reason than that cash
was scarce and drovers would extend credit for the year after
new cows entered herds.

Because times were changing, though, those who stuck by
their guns failed to prosper. Yesterday's smart farmer found himself suddenly old-fashioned. It must have been bewildering and
demoralizing. Buying from drovers had once been a most sensible practice; farmers wanted to see what they were getting, they
wanted to avoid breeding bull-calves, which give no return for
time spent tending mother, and they wanted to avoid spending
money for the labor, housing and feeding of new heifer calves
for two or three years before they joined the milking string,
only to find that some new milkers had no talent for their jobs.
They wished not to tie up scarce capital awaiting an uncertain
outcome; better, they thought, to spend it on a known quality
of animal such as they could see passing before their farms in
the drover's herd.

Yet soon after the questioning of the old ways began, the
advocates of farm-bred cattle had won, and the reason was that
the full-fledged dairy industry had come of age, changing for
the first time in millennia the reason why people kept cows. The
proponents of drovers' replacements were not wrong so much
as they were slow to perceive the changes of the organizational
structure already underway. By 1877 — just a decade after
Thomas — X. A. Willard, writing in his classic *Practical Dairy
Husbandry*, states with certainty:

No one will deny that one essential requisite to success in
dairying is to have good cows for the business, and in considering how this is to be obtained, it is evident the surest
method would be to rely upon breeding animals upon the farm

where they are used, rather than to purchase at random from
droves. . . . The only practical course . . . for the majority of
dairymen is, to start with a good herd of native stock, using a
thorough-bred bull, and breed up to the qualities desired.

The race was on. The cause of the sudden change in correct
farm management tactics is well rehearsed: factories grew as
transportation improved and the cities grew around them. City
people drank milk. The demand by farmers for more efficient
milkers grew too.

Demand for new technology by farmers is constant, but does
not come simultaneously from all practitioners. Farmers as a
group alter their business tactics slowly. The very first to
perceive the winds of change may in fact be too speedy to
profit most. At first, refined breeding of dairy cattle — before
there was a well-formed milk market and before development of
a structure of record keeping allowed sale of fine bulls at high
prices — was a gentleman's hobby. Farmers early, but not
pioneering, in adopting new technology are best reimbursed for
their slight daring and orderly perception. Late changers-over to
technology that has become widely adopted change in order to
keep from going broke. The switchover to farm breeding took
place over the course of a generation — Ray Totman remembers
drovers still stopping at the farm when he was a young adult.
The transition had several stages.

At first, neighboring farmers shared good breeding stock for
compensation. Then farms developed that specialized in sale
not of milk but of dairy breeding animals. And for several
decades before World War Two, regional companies extracted
semen from their bull studs, refrigerated it and shipped it on to
customers. Good roads and air travel increased the scope and
competitiveness of these regional suppliers of genetic material.
Then, in the late 1940s, an accidental invention — the discovery
that semen mixed with glycerol can be frozen nearly to absolute
zero and preserved for many years — again transformed the

dairy breeding industry. The semen of a top bull, diluted and frozen, could now service tens of thousands of cows in their own barns. Fewer bulls than ever before were needed, and competition to provide the highest-quality semen possible grew fierce. In this high-stakes commercial climate, only the best funded of companies survived. It was as if the companies themselves had adopted the combative nature of the beasts whose genes they spread.

In 1953, according to an article in *Hoard's Dairyman*, when this commercial struggle was well along already, there were about a hundred semen-producing businesses in the country. Nowadays only twenty remain. The hundred companies inseminated five million head a year. The twenty inseminate eight million head. The cost of developing or purchasing top bulls became prohibitive to all but the largest companies and farm cooperatives.

The net impulse of the improved breeding technology, to reduce greatly the number of dairy cows needed to feed Americans, has concentrated within a small number of companies the profits of "breeding up" that used to be widely dispersed. Money that once went to drovers, and later to whatever neighbor had a sturdy bull to lend, and then to local and regional insemination companies, now travels to national organizations owned, in many cases, by still larger conglomerates. Because it costs a farmer more to keep a bull on the farm than to purchase semen from one of the twenty breeding companies, the huge stud services thrive.

The success of the Totman herd in this demanding business climate has been the success of a smart family that managed to turn to its advantage inventions that drove most Yankees out of farming altogether. Not the least of their successes has been their own attention to "breeding up." Frederick kept a good bull; Raymond began to use artificial insemination even before the invention of deep-frozen semen made the practice widespread. And Leland, a loyal customer of Eastern Breeders

Service, is a student of the top bulls available from one of the large stud services. He keeps his animals extraordinary because it pays. The first eleven or twelve thousand pounds of milk a cow gives meet the costs of farming. The profit over costs increases with each subsequent ton up to the level of Lee's herd average. Lee balances where he wants to be — riding the crest of the wave of modern breeding technology.

As wondrous as Lee Totman's herd average is, as fine as his top producers are, milking 25,000 and even on occasion 30,000 pounds of milk a year, "or at least a few of them could, I guess, if I babied them the way those fancy breeders do, feeding them six times a day in individual stalls," as wondrous as a 19,000-pound herd average is, it scarcely gives a hint of the freakish genetic potential to make milk that has been bred into some special dairy animals.

There are, according to *Hoard's Dairyman*, still cows at work in Ethiopia that seem to have little "breeding." They yield a thousand pounds of milk a year. But in Rochester, Indiana, in a box stall, in a Quonset barn, with adjacent paddock replete with well-nibbled apple trees, there dwell three mascot sheep, several friendly geese, and the queen, the reigning bovine, the Cleopatra of all dairy cows, whose name is Beecher Arlinda Ellen. Queen Ellen is the Roger Bannister of dairydom. She has broken the four-minute mile for milk production during a measured three-hundred-five-day period. She has broken the 50,000-pound barrier with 314 pounds to spare. That's more than 24,000 quarts of milk from one cow — Lee's fine cows average just over 9,000 quarts a year. Beecher Arlinda Ellen, at the top of her production, eats 60 pounds a day of 16 percent protein grain, and another 70 pounds of alfalfa hay. She drinks nearly 60 gallons of water a day.

Ellen is, these days, nursed and tended with care. Her proprietor, Harold L. Beecher, says, "No water ever fell on her back." Her record milk production earned him about $5,000. Her progeny, male and female alike, are enormously valuable breeding stock, worth tens of times the value of her milk. The shadow of Beecher Arlinda Ellen darkens the ruminations of somber and ambitious farmers the nation over. But for all her worthiness, can she sit on velvet cushions on a high throne, and languidly munch hay cubes? Can she snooze, munch kudzu, romp with young calves and young bulls, leading games of follow the leader?

Cows, even lesser cows, exhibit their competence to exist these days by going about their business continually. Dairy cows' business is eating. The wild cows of yore also had to find food and keep themselves out of danger. But with humans in charge, the food is delivered and the danger institutionalized.

Eating's the only job left, and cows must go about it with a will. All animals, of course, eat and turn food into flesh. Lactating mammals turn food into milk as well, and dairy cows, having been bred to the chore, replicate Beecher Arlinda Ellen's hunger. Beecher Arlinda Ellen is the queen of hunger. She's too hungry to loaf. She's condemned by her genes to a lifetime of insatiable hunger. She was born to crave food, to feed, to metabolize food and make milk of it at a rate that fascinates dairy scientists, and to eat more while they stand about fascinated. She eats instead of sleeping.

Beecher Arlinda Ellen eats while cartels compete for the honor of buying her calves still damp. Her daughters will be raised to test and compete with Mother. Her sons will remove to fine pastures and will soon enough be at stud. The stakes in these subsidiary Beecher Arlinda Ellen enterprises, like the bureaucracies needed to sustain the complex dairy breeding industry, are enormous. The breeding industry's procedures are costly and complex, the results quixotic. Beecher Arlinda Ellen keeps on eating.

Beecher Arlinda Ellen represents the pinnacle of accomplishment in the vast and dispersed effort dairy farmers engage in to put each other out of business. The horsy set, whose goal is to push galloping time for the measured mile ever closer to the speed of light, are a mild and amiable lot compared with cow breeders. Horse breeders who talk about "the good of the breed" dream of fair lines and of whupping their neighbor in the Preakness — and perhaps are delighting in the sort of equestri-chauvinism that allows the family-proud and bone-weary to participate vicariously in the demonstrable genetic vigor of their hobby horses.

How different it is with dairy cow breeders. They too are seekers after "goodness of the breed," but only in passing does cow breeding seem to have attracted julep-misted seekers after ancestral blood. What successful dairy breeders have to sell is an economic edge over their neighbors, and a way of "milking" more money from *Bōs typicus*. It is different from what makes horseraces. The accelerating pace of dairy breed improvement makes more farmers poor than it makes rich. It serves to concentrate wealth in fewer and fewer farmers' pockets. It puts poorer farmers out of business. It forces all farmers, regardless of inclination, to participate in a mad race to keep up with the Beechers.

The dairy breeding industry is built on the work of a minority of farmers who strive to make finer genes as well as fuller tankfuls of milk. Eight-five percent of dairy animals in America are Holstein-Friesians. There are about nine and a half million of what are called "grade" Holsteins, which are pureblooded cows but without pedigree papers. Lee Totman keeps a "grade" herd. A twentieth of Holstein cows — about half-a-million animals — are "registered," nearly all with the Holstein-Friesian Association of North America. The infrastructure supporting the HFAA demonstrates the complexity, and still more, the structural, unplanned, automatic, business-as-usual nature of the forces that make farming today a race for technological improvement.

To start with the half-a-million registered cows: at a cost to a farmer of perhaps thirty dollars a year the cow, an inspector from the Dairy Herd Improvement Association (a testing service run by the Department of Agriculture) shows up in farmers' barns twice yearly to record, with unimpeachable honesty, milk yields of each cow. From these tests are extrapolated yearly yield averages (many "grade" cows, including Lee Totman's, are on DHIA test programs also, because the results guide efficient feeding and culling choices).

Farmers owning registered animals file sketches, bloodlines, and registry numbers for each cow with the Holstein-Friesian Association of America, Inc. These farmers are basically interested in coming up with choice bull studs whose daughters will milk more than the daughters of other bulls. They are also occasionally interested in breeding not only for production but for "type" — for daughters whose udders are higher, whose hooves are sturdier, whose faces are nobler of aspect, who conceive more easily and calve more easily, who resist disease more heartily and stand more proudly than do the daughters of other bulls. But mostly, they are interested in breeding bulls whose daughters make more milk.

It is not easy to establish reliably whose daughters do these things. One must raise up a purebred bull until breeding age (about two years), breed it artificially to sixty or eighty daughters in scattered herds, await the birth of thirty or forty female calves (that takes about ten more months), raise the calves to join their respective milking herds (another two years) and then milk them for a season (ten more months) — it's six years before the results are in. This is costly — it is estimated that it costs at least forty thousand dollars a bull. But thousands of novice bulls are aspirants each year for the role of father superior. The motive of farmers for engaging in bull testing is the chance for profit, and the joy of speculating on a prideful bit of uncertainty amidst the plodding routines of daily farm life. The motive for the Holstein-Friesian Association of America is fiscal soundness, and, as their publications benignly put it, "the good of the breed." And it's true. Their hand is forced too. If they let up, Brown Swiss, Jersey, or Guernsey will forge ahead.

HFAA certifies and keeps records of the winners, those rare beasts whose "predicted difference" — a statistical index of how much a bull might raise a farmer's herd average — for both milk and type is high. HFAA is at the crimp in the information flow; it performs the massive quantity of secretarial service needed to isolate the fancy bulls reliably.

The mother church of Holstein breeding is a big brick factory building on the main street of Brattleboro, Vermont. The view from the office of Robert Rumler, executive chairman of the Holstein-Friesian Association of America, is spectacular — he glances out the window now, down across a railbed and the flats of the Connecticut River, across the broad stream, up the logged-off slopes of the far bank, where Yankee cows of some earlier and less certain breed must have scavenged their suppers. Rumler is a powerful man — one of those mysterious people who are in charge of something vast that most people have never heard about. He controls the genes of America's milk supply. He has built the power of his organization from the time it first emerged as the dominant force in dairy genetics, has shaped it into a great bureaucracy that can dispense reliable information about whose good cow is which, and when it did what with whom. Because he has built a powerful position, he flies around a lot. His testimony is solicited in industry, commerce, and in the halls of Congress. He orates about foreign market development funds, and about the world food crisis.

Rumler's major difficulty is keeping HFAA preeminent. HFAA does its job ploddingly, thoroughly and on a massive scale. That is its commercial advantage. It turns out that there are plenty of good bulls, but, by definition, few fine ones. The ponderous system set up to isolate those fine animals from a rather large population of excellent candidates is costly. That the system works, that it is consolidated, stable, reliable and beyond question is an exhibit of the dull and relentless power of bureaucracy to shape the economic conditions of peoples' lives. No one appreciates the slow power of the legions of silent file clerks and functionaries better than Robert Rumler. In the Brattleboro office millions of cattle facts are recorded: births, names, lineage, verified with photos and sketches of patterns of spots, blood factors taken and cross-checked, records of milking performances, breeding performances, success in transmitting traits, and dangers of transmitting flaws.

What seems miraculous is that a cooperative effort requiring broad funding exists even though few benefit and many participants are harmed by the results. Most farmers keeping registered Holstein herds make little from the fact of their cows' registration. HFAA is a drab empire built on small farmers' dreams. It works to the detriment of most of the dreamers.

The power of the HFAA derives from the act of filing information, and so the organization is rather unusual as offices go — consisting of a very few executives and of very many file clerks, some of whom have been recording cow numbers, looking up cow grandmothers' numbers in banks of file cards, sketching cows' spots on standardized grid forms for stretches of forty and fifty years. The office filing pool, which seems to imitate the bovine disposition favoring routine, works in a large hall under a large-as-life oil painting of "True Type Holstein, Male."

A model of his ideal mate, "Ideal Type Holstein, Female," sits next to another statue — of a jaguar about to spring — in Rumler's office. "Ideal Type Holstein, Female," is large of barrel, high of udder, sturdy and shapely of leg, and straight of back, with an all-around "dairy-type" face. She has never existed, and if ever a cow is born that resembles her, she will be immediately put out to ideal-type pasture and replaced by a cow of still more unattainable perfection. It's a case of carrot on stick. The jaguar statue guards the cow statue. The fierce cat crouches coiled to spring from the corner of the river-view window ledge down upon the adversaries of increased production per cow, whoever they might be.

Rumler greets me from behind his desk. He peers over heavy black-frame glasses and wears a white gabardine suit. He lays a trap for the interviewer. "I suppose you're concerned about the takeover of American dairy farms by large corporations, aren't you?" he opens, smiling at me.

I give the right response, dutifully. "No, sir, it seems to me more complex than that. The USDA finds a small percentage of farms actually owned by large corporations, although there are many closely held family corporations that are actually family farms in modern dress. Corporate farms have had more success in vegetable and grain and poultry businesses than in the dairy marketplace."

"Very good." Rumler says. His desk is bare. "Now what is it you want to know?" He turns out to ask this question to take in my answer, not to oblige by coming up with the responses. He answers questions like a reverse Chairman Mao — elliptically, allegorically, with figurative generalities that are useless to quote out of context. He used to be a public relations man for Du Pont. He gives enthusiastic speeches with such titles as "Applying the Formula for Success." He says things such as "Success is a journey, not a destination."

The problem is, I suggest to him, that at this stage of genetic advancement, what is good for the ruling powers of the HFAA may not be what is good for either America's dairy farmers or dairy consumers. I ask him about the "apartheid" policy that keeps excellent-grade Holsteins out of the marketplace in breeding stock, confining that lucrative business to the few farmers whose families have kept up their cow registrations or who have bought into registered herds. I ask if there isn't a huge grade gene pool that ought to be tapped.

"I don't want to be hedging," he replies, "but I don't know if anyone *knows*. Geneticists would say, 'Yes, there is.' To add variability is to make more rapid improvement. In this sense, any broadening of the gene pool leads to increased genetic capability."

The HFAA has recently started a program for identifying grade milkers, but has not started one for testing grade bulls. There seems little chance in the near future for the legions of grade dairymen to cash in on their years of breeding labor and investment.

In the meanwhile, dairy farmers persist in dreaming about the new improved cow. "Think of what my milk check would be," farmers think, "if my cows gave twice as much milk each day." If all cows gave twice as much milk, the check would be about the same, and business would be tougher. The profit of a high-producing cow isn't merely in its productiveness, but in the rareness, the timeliness of its high production. When other cows catch up, that's how good cows will have to be simply to avoid the butcher's block. And cows are improving rapidly. The public wants only so much milk. But the national herd average is improving at an amazing three or four hundred pounds a year these days, after decades of slow improvement and centuries of no improvement.

For all this rapid change, one would be in error to consider the HFAA a big business conspiracy. It is just one key part of the great engine of progress, insensible, elaborately organized to foster change, and containing no mechanism for considering the consequences of that change.

And there are consequences. Because cows are better, more money can be spent on keeping them and milking them. It pays well to buy the best stall for a cow to sleep in, the best system to milk a cow, the best feed to keep cows' production rolling, and the best semen for breeding herd replacements. All this means more business, more financed agriculture, more low-labor, capital-intensive farming systems, and ever fewer farmers.

There are two ways to look at dairy progress, and as with so many elements of modern technological improvement, it seems impossible not to hold both attitudes at the same time. On the one hand, improving dairy genetics and hardware are newly parsimonious and therefore wondrous — it is an aesthetic pleasure to see the sleek operations of the farmers such as Lee Totman, who have managed to stay in business. On the other hand, it is clear that modern farming machinery and husbandry — and the capitalist system that has urged these things upon us — have rid us not only of drudgery, but of a rewarding way of life, and have replaced it not with a superior one, but with one more isolated, dependent more on forces beyond a farmer's control, one lacking the local culture that once made so much sense that most farmers never thought about how things *ought* to be at all.

If the present has left farmers bereft of their culture, and if some of the transformations in dairy farming that have had this effect are attributable to "progress" in dairy genetics, the future is still less promising. Two new lines of endeavor, currently the subject of intense research by scientists in both private and public sectors, will almost certainly alter the world of dairy farming suddenly and strongly when they come of age. It is probable that they will develop soon, and certain that when they do, the dislocations seen so far in the changing nature of dairy farming in New England will be replicated on a still grander scale.

The first of these developments is semen sexing. Scientists are working to develop a process that will enable stud companies to sell semen guaranteed to produce offspring of predetermined sex. Of course all farmers not engaged in breeding stud bulls will always choose to have their cows spring heifer calves. Sexed semen will double the efficiency of calf raising, since now half of a farmer's calves are not of the desired sex. With more heifer calves to choose from, one may expect herd averages to climb, decreasing still further the total population of cows — and of dairy farmers — needed by the nation. It will be delightful to the farmer in the short run to have such control. In the long run, such increases in efficiency, given a fixed demand in the marketplace, inevitably result in a continual shakeout of "less efficient" producers.

The more Faustian and far-reaching scientific innovation, now far along in its development, is that of ova transplants. When perfected, ova transplants may result in as sudden and significant a rise in dairy production per cow as that caused by the introduction in the early 1950s of frozen bull semen from top bulls. Ova transplants are now done by a surgical process in which a veterinarian flushes a cow's uterus to recover eggs recently inseminated. The uterus flushed is not the uterus of any cow, but of a "top cow," perhaps of Beecher Arlinda Ellen, or one of her sisters. The process releases up to twenty fertile eggs. Awaiting their blessed events in a nearby barn will be twenty guest mothers, their estrous cycles synchronized by drugs with that of the donor cow. The recipients are implanted with the fertilized eggs — this used to be a surgical procedure, too, but recently has been accomplished without surgery — and grow up healthy calves. The calves, a score at a time, are all prized stock. In the future, a farmer may be able to take fertile eggs from his best cow and have a barnful of her daughters at a single throw. Needless to say, it would result in quantum leaps in milk production in every herd that used it. If the technology is patentable, it may place a corporation in a position to intervene in the determination of the value of efficient cows.

And, further in the future, what will happen to farmers when Beecher Arlinda Ellen can be cloned?

III

A nother year's round of farming is over. The farm goes on. Raymond is back in his easy chair, reclining by the dining room window that overlooks the dooryard. He spends more time there than he did a fall earlier. His legs "took up lame" a few months back, and he has moved from two legs to three, and then four, first using a cane, and then crutches, to get around the house. He can no longer mow hay. Mildred clatters about the kitchen. She is as strong as ever. Raymond's older sister, Ruth, lanky and wiry in her mid-eighties, who much of her working life taught physical education at the University of Massachusetts, helps Mildred. They chop and stir and taste and mix, preparing a picnic. The family is converging from all over. It is to be a harvest celebration. Lee has recently capped the silo after loading in the last of the corn harvest. He has confined the cows for the winter; they are eating silage and grain. He comes and goes from the yard now, doing small jobs, and Raymond never fails to observe through the window his son's progress.

If his lameness has changed him at all, it has left Ray fiercer and brighter than ever, making up with force of thought for the want of strength in his body. He is in the mood to take measure of the changes in the world during his long life.

"I do think," he says reflectively, "that I got more pleasure out of my lifetime of work than I'd guess Leland gets from his work, although you can't really say with things like that. The new technology takes fewer man-hours — modern technology does save much manual labor. But it has liabilities.

"The noise will make a young man deaf. The tension of operating the equipment is so great that the recovery of a night's sleep may not be enough. It's tension that makes you tired, you know, not muscular exertion. Lee mostly works at things alone, and I worked at them mostly with at least one other person or with a team of horses. That wasn't so bad, either.

"The horses rested and you had to rest, too. And plowing with horses — they knew where they were going, what they were going to do. You were freer. The mind was able to soar, engage in flights of fancy or deep theological meditations." When Raymond indulges in irony, he never smiles, but his eyes grow round, giving him an impish appearance.

"My neighbor Will Graves — Lyman was my contemporary and Will was his dad — Will was not a poet, but he put together doggerel. He'd make them up as he worked with the horses. He never put those poems to paper. His sons are gone too, and his generation is pretty well forgotten. I just mention it to show that it was different with horses. I remember one — it was when my brother Bill was here farming, might have been nineteen nineteen, and my sister Mary brought over another schoolteacher to stay with her . . . no I can't tell you that poem, and I'm sorry I mentioned it, there's no persuading me to do so." His eyes grow still rounder.

"It was always more pleasant working with other people. Some of the hard jobs were the most pleasant. Using a crosscut saw was an art for two people. Cutting trees so slowly, they had more chance to split. It was a real source of danger. You had to work well together. Once, after the great hurricane, there were tipped trees. We cut out a sixteen-foot-long sawlog from the top end of one and the tree snapped right up straight again.

"For tobacco, it took a certain-sized crew of neighbors to be efficient. The plant had to be suckered, cut, wilted, then put in, hung on crosspieces. A camaraderie developed if they were a congenial bunch, which mostly they were. The drudgery and fatigue didn't hit you so hard if you were a group. Young folks handed the ties up, loaded with tobacco, and by four in the afternoon they were tired. So we'd sing to them. We'd discuss the latest show. We'd tell them stories, tease them, anything to distract them from their fatigue. In midafternoon, the womenfolk would bring out cornmeal johnnycake with butter, and coffee, and we'd all stop and visit for a few minutes. Those jobs were hard, but pleasant.

"I caught the tail end of animal energy, beginning of motor energy. For the animal part, though, I was young and in the prime of life. When it was more mechanized I was tired — maybe that accounts for the difference in pleasure."

Mildred and Ruth come and go through the door, and the banquet assembles on the patio. Chicken and chipped beef in thick white gravy, roast beef, coleslaw, fruit salad, potatoes, stacks of home-made rolls, carrot breads, pies, raspberry tarts from the last of the year's berries, applesauce from the trees behind the house, cupcakes, a large pitcher of milk. The generations assemble, Lee, his wife, Betty, Connie, Gail, Barbette, "the children," meaning the eight grandchildren — including diverse future professors and farmers and also a tall teenager named Val who has arrived back from a pensive hike through the family woodland, a folded copy of Gorki's *Lower Depths* tucked under her arm. Raymond stays indoors as we line up for the feast.

Lee's brother, Conrad, and I pile food on our plates and sit together with Lee in the fall afternoon sun. We eat silently for a while. I think of my previous meal and discussion with Leland; this time we talk only of cows and of the good weather.

Connie, the professor saved by mechanization from a lifetime in the barn shoveling manure, is a slight and scholarly man. He looks professorial now, even in his torn house-painting pants. Lee is soon finished and goes back to working in the dooryard, quipping with Connie on the patio whenever he comes within earshot. Raymond has made his way out and now sits at the end of the banquet table, eating lightly but obviously enjoying the gathering. Mildred slices and serves pie, then, finally, sits with us.

"There are only two times in my life that I ever thought about moving away from here." Raymond gestures about the farm. "One time was when I went off to college — but I was back within a year. The other was after the barn burned down."

"It's about the worst thing that can happen on a farm," says Mildred. She sets a cup of tea in front of each of us.

Raymond sips his tea, thinks back. "Firefighting's a good way to kick off if you have a bad heart. It's one of the forces of nature that makes man realize his insignificance. The barn was hit by lightning. We were haying, it was the middle of the first cutting, June, nineteen hundred and forty-six. The cows were out to pasture — quite far from the barn, fortunately — but half the first hay was already in and we had a few pigs, chickens, and an old bull quartered way in the back. We were haying and the sky got dark — a summer storm. We'd seen it coming, worked until the last minute, then put away the horses and ran for the house. I dried off, sat down, dozed a bit."

Mildred takes over her part of the story. "I was looking out the window at the storm, and then I saw one bolt of lightning come right down at the barn. 'That last one hit here,' I said. . . ."

"I didn't believe it," Ray says, shaking his head still. "I dozed again, then she says, 'There's smoke coming out of the side window of the barn.' We ran out, took out the horses. I was kicked by one because I was too excited. We took out the milking equipment, pushed out the manure spreader. We went for the tractor next, but when we opened the doors, the heat drove us away. It was too great. After the fire we found the tractor two floors down. The bull couldn't be reached, and we lost him, also a pig, beehives out back, sheltered behind the south wall of the barn, and some chickens."

"The chickens were mine," Connie says. "I'd forgotten them — a 4-H project, I believe. I was fourteen. I still recall looking in through a window at those chickens. I saw them panic and take off from their roost, trying to fly. As each bird rose into the air, it burst into flames. . . ."

"I had a pile of sawdust for bedding," Raymond takes over again, "and it burned and smoldered all the day after. The fire laid the building flat. The fire department finally came back and drenched it. Still, it would start to flame up. We'd douse it with water. It was weeks before I could sleep without waking up and thinking I smelled smoke.

"Now here's the odd thing: a man had been talking lightning rods to me not too many weeks before. When we rebuilt, we rodded everything. The children were as apprehensive as the dickens from the experience, and we used the rods to lighten their fears."

"There is one photograph of the old barn in the family scrapbook," says Mildred. She goes in and returns with the picture and we take turns peering at it — just an average, practical, multitiered, built-piecemeal Yankee barn.

"After it burned, it took a week for us to recover heart and plan for the future. All that time we kept milking the herd — used a neighbor's barn one day, then set up a shed and milked here. But I knew all along what he'd decide," Mildred says. Although she is reluctant to admit it, it was her thought at this point to buy the adjacent "lower farm," which had been last united with the "upper farm" a century earlier. This vision has proved a key to the continuing success of the family operation in modern times. There is still land to grow into.

"We built a new barn before summer was over," she concludes.

"I'm glad we stayed in it," says Raymond. "I can see what I've done here. Maybe you can call it a fiefdom." Raymond looks at his large family. "To have continuity of this particular little bunch of toadstools I call my domain was very important to me. In fact, I'd say my son taking over was one of the main goals of my lifetime. I'd wanted both sons in, but now I'm sure it's damned lucky it didn't work out that way. I think they are both happier the way things have turned out. This is the place where I was born, the place where I grew up. It carries adolescent memories, a lifetime of experiences." Mildred clears her throat.

"Perhaps some of the motive for wanting Lee to take over is selfish too," she says. "There's no prettier place — if Lee didn't take over we'd have to move along about now. It's such a lovely spot. The children are attached to it. It makes them glad to come back. They all have their little chunks of land here." Raymond turns to Conrad suddenly and asks, "You know what the most endangered species of all is?" Conrad smiles and shakes his head slowly from side to side.

"The most endangered species there is today is man himself. Not atomic power — famine. Without oil we can't go on farming like we do. It will stop.

"After that it will be survival of the fittest. It could be nasty. This family won't necessarily have an advantage there just because it has a farm. When there's a shortage, the strongest or most cunning will take it. Reason treats man fine when most everyone has a full belly and a comfortable place to sleep. But the hungry person isn't going to be stopped by a Mahatma Gandhi, or a President Carter.

"But even so there will be continuity. Just look out-of-doors." Raymond Totman makes an encompassing sweep with his arm. "See the grass growing up in the driveway. See the abandoned fields on that sidehill growing up to forest. You realize that unless there's some terrible catastrophe, water will run, and grass will grow, and streams will flow on over. . . ."

Family Farmer

J oe Weisshaar's wife says she was first attracted to him because he was *lowly*. It's a modest agrarian virtue, an ask-no-favors, sturdy, bullish, and Rotarian thing to be, a fine, German-Catholic, almost Christmassy trait. "I was impressed when he first started to pay attention to me," Mary Jane Weisshaar explains, "by how humble, how *lowly* he was in church." Joe's off-duty expression — the one he wears when he is waiting around for a moment while his son, Al, does his part of a chore, is rather mild. It's a shy, slight smile.

He looks like a tamed gypsy, wavy brown hair never quite as kempt as it ought to be, a swarthy, weathered complexion, and deep-set eyes that remain alert even in those off-duty moments. Because of the smile and because he is of uniform thickness from head to toe, like the short trunk of a tall tree, Joe seems to be moving slowly even when he is moving right along.

His mother, who lives down the road, recalls that Joe was a football star in high school; the conservative editor of the local newspaper insists Joe "was not a star — just average." The editor also says Joe is "a known Democrat. Very active type."

Joe is assertive. His "lowliness" is evidenced not by any discomfiting self-effacement, but by his soft style — it's an athlete's evenhanded fairness. When as a boy, Joe played backyard ball, his friends must have asked him to be captain.

The principle of his adult assertiveness is survival in a nasty world — survival not just for himself (one gets the feeling that were Joe a bachelor he'd eat from packages and wear his socks a day too long) but for his family, and for the family farm. Joe is very much the family farmer.

"Nixon's Secretary of Agriculture, Earl Butz," says Joe, combat-ready, "he came out and said farming is just a business nowadays and family farming better face it. It's not just a business. It is still a way of life. There's always something to do instead of drifting aimlessly around town." Joe has a *farm* and he has *values* — the two go hand in hand in his life and both are under siege.

The Weisshaar farm appears at first glance to be rubble-strewn, disorganized, haphazard. The house is not one of those long, low ranches that all through the Corn Belt indicate that farmers are *arrivés* enough to imitate suburbanites. It is long, but also high, old, complicated, roofed in two separate styles of shingling, one red, one blue. "Their home is a nice basic home," said the conservative local newspaper editor, "nothing elegant." The newspaper editor had a ceramic model of a golf bag holding pencils on his desk.

A false start of a garden has yielded a few pale tomato plants dwarfed in a grove of weeds before the house. The barnyard — staging area for all on-farm action — is in a state of barely controlled chaos. A long hog house whose cement block walls have cracked sits off at the east end of the yard. Four supplementary outdoor hog pens, black, rooted ground surrounded by mesh fences, small sheet-metal shelters and large battered orange self-feeders in every one, show that the operation has expanded since the hog house was built. At the northern edge of the barnyard, a white hay barn that needs painting protects a couple of dozen yearling steers from the stormy fall. Worksheds are strewn all over the lot. Weeds ring every building. An old grain bin has been split into two half-cylinders and both halves lie near the house, garaging machinery and trucks.

There are two schools of neighborhood gossip in Iowa about messy barnyards. One school has it that a messy yard betokens a messy business. The other, more tolerant of disarray, allows that a farmer may be more interested in getting things done than in prettying up after himself, and that while tumult in the barnyard might indicate sloth, it also might signal a farmer going flat out, who hasn't found a moment recently to run for the broom.

Joe tends toward the second sort. With help from his wife, one lethargic farmhand, his son Al, who is eleven, and a large complement of fairly new machinery, Joe Weisshaar farms about five hundred acres of rolling, rather dry hill land in Creston, Iowa, in the southwestern corner of the state. On the farm grow good pasture, good hay, soybeans and corn. Joe feeds most of his crops to hundreds of head of beef cattle and thousands of young hogs. "The hogs are the main thing," he explains; "the cattle are here to eat the roughage."

Joe's is the first farm out from the center of Creston, just past the junior college and the golf course. ("My pickup's been hit twice this year alone. Last time, the ball went home in my grill.") Like most of Iowa, the Creston area is not the bleak, flat country of the high plains. Joe's farm has virtually no flat land on it, although a few lucky Creston farmers do work "first bottom" pieces along the creekbeds of the town. Joe regrets the rolling hills. A family story about flat land explains his feelings.

"In nineteen eight," says Joe, "Grandfather moved out of Tainter, Iowa, and came on down here. I think he wanted a more active church than he could find way out where they were farming. Story goes he looked at some real good land over in Marshalltown — land's richer up there — but he didn't care for it. Said if there's been one single acre higher than any other so he could build a house on it he'd of settled there. He came down to Creston to farm. Liked the hills. Oh, how I wish he'd stayed on that flat piece," says Joe longingly.

Joe is interested in good land. He is hurting financially, and for modern large-scale farmers, the way past financial troubles is further expansion. Since 1950 the average Iowa farmstead has increased its acreage by more than 50 percent. To survive is to get bigger. Growth, in farming, comes via a sort of economic rocking motion. Bigger equipment increases the acreage, bulk of crop, number of animals that each farmer can handle. The big rigs can cover more ground; and they also demand more ground to amortize their high cost.

It is actually a wonder that Joe is looking for more land, because for three generations, the men of his family seem to have made wrong decisions about acquiring more land — favoring instead at every turn the use of spare cash to buy more machinery and better stock, while they continued to rent most of the land they farmed.

Even today Joe owns, together with the bank, only a hundred sixty acres — less than a third of the land he farms. He is a minority shareholder on another couple of hundred acres of land owned principally by his uncle. He farms that land, and a few other "pieces" around town, on "shares" — which means that at the end of the year he gets half of his harvest and the landlord gets the other half.

This has been a bad year for Joe and the other farmers around Creston. It was droughty through mid-August, crippling the corn crop, stopping hay and pasture growth altogether. The soybeans grew stunted and "unfilled," until the onset of a great late-summer deluge. It rained from mid-August on, but it came too late for the corn. It rained with such frequency after August 12 that there was never enough time to dry late hay, and all the hay that Joe cut after the ending of the drought was rained on. The soybeans had hung on well, and the same rains caused the empty pods to fill. Joe expected a nearly normal harvest.

But it never stopped raining long enough to let the fields dry.

The heavy combine — aluminum, boxy, and big as a cottage — sat greased and ready in the yard. "Three days is all the harvesting I've gotten done this fall," Joe says with the practiced weariness of someone who has sat with disaster for so many awful days in a row he is used to it. "I usually go flat out from mid-September until after Thanksgiving."

In the face of this disastrous harvest season, Joe is hunting for more land, for more cropland and more grazing. This may seem contradictory, but in modern farming conditions, hard times frequently suggest the need to expand — to get more work out of expensive equipment, to use the full time of a hired hand, or a maturing farm child's labor, to combine more acres farmed at small profit, to take advantage of land freed for purchase by neighboring farmers' misfortunes. Unlike the past generations of Weisshaar men, who buttoned up and waited things out, Joe is burning with desire to look out for his own. He's a businessman as well as a farmer. And in business he has had some good fortune recently.

The land he does own has, along with the rest of the farmland in the Corn Belt, greatly increased in value in recent years. The average acre of Iowa farmland that sold for $88 in 1940 and for $257 in 1960, today is worth about $1,500. Joe's net worth has ridden upward correspondingly. He has parlayed the increase into increased borrowing capacity at the local bank. The credit line has helped Joe farm, and it has put him in the land market.

In the pursuit of familial duty, he has done some good business around town as well, developing and renting out several small offices. And he fell into a windfall purchase of a field full of government surplus grain bins, abandoned by Earl Butz. Joe rents crop storage facilities to his neighbors now, and it pays for family groceries, even in bad years.

We drive away in the family pickup, right after morning chores, heading out across town to fetch the old fast-dealer who sells real estate. Joe doesn't like real estate brokers, he tells me as we drive through downtown Creston. "They started coming around my dad's place when the poor fellow was sick but still living." The broker used to farm but didn't make it. In fact, for a few years, he farmed land that Joe farms now. The fellow clambers up into the truck. He is creased neatly, doubleknit angling sharply over pipe legs, his face articulated and warmed by the well-marked sunrays of a practiced smile. He's got a cupcake of a chin. He smells of cologne and chews gum.

We come to the land. Joe steers the four-wheel-drive in through pasture gates. We bump over rough ground, dip through grass-floored waterways, slide down washing and muddy hills into a steep-sided field, climb up the far bank and come out onto a rolling cornpiece. It is the best ground on the farm, although the best here is none too good. The fields show signs of long neglect. The hayland is weedy, the cropland rough. Lines of old corn stubble ride right up and down an eroding hillside.

Joe says: "It's not so rough but you could farm it."

That's a midwestern way of talking badly about what the other fellow has to sell without ceasing to be polite — it's part of the region's code of honor, and all follow it, but all say what needs to be said anyway.

The broker says: "There'd be some good building lots. You have a few nice houselots in here, too."

Joe says: "I wish you guys'd stop pushing the houselots out here — there won't be any land to farm soon."

And the broker says: "I agree. I never do that."

It's welcome to the Midwest.

Joe mentions to a farmer we visit that I've come out from New England, and as if to make me feel at home the farmer says — while pointing about his eight hundred acres of fields —"We have a few rocks one place here."

As we bank down through the first flurry of the winter into the airport, a fat, fiftyish Marine lifer jerks a beefy thumb past my chest, gesturing at the view from the plane window. The horizon forms a diagonal slash — twilight on one side, dark ground on the other.

"There's your good farmland," he says.

And he's right. A quarter of all the land the USDA calls "class one" is in the state of Iowa. So are 135,000 farmers (Massachusetts has only 4,000 farmers left). They farm 95 percent of the land within the state boundaries, some 34,000,000 acres of land (Massachusetts farmers till only a quarter of a million acres). In all the United States, Iowa ranked, in 1977, number one in production of corn for silage, number two in corn for grain, number two in soybeans, number three in oats, number four in hay. It ranks number one in hog production, number two in cattle, number seven in dairy animals, and, for that matter, in turkeys. It is tenth in numbers both of sheep and of mink. A third of Iowa workers are directly involved in farming, while most of the rest are involved in supplying farmers, in processing what farmers grow, or in hauling something to do with agriculture into, out of, or around the agricultural state of Iowa. Iowa ranks second in the nation, just behind California, in total value of agricultural marketing. But a still greater industry in the state than the growing of food is the handling of it. It is Iowa's number one source of income.

It is a wealthy state. The income of Iowa's farmers for the past five years averages about $11,000 per year. Nearly everyone in the region has to do with growing things. Those not still farming can trace a lineage to farmers at the most a generation back. In that sense, Iowa is a rural state. Farming is the news. Strollers on the streets, even the streets of Des Moines, talk farming. When the politicians of the state make political speeches, they are always speeches about farmers and how to solve their problems.

In another sense, though, Iowa is most unrural. It is a state that has been tamed. New England (which during one mad score of years around 1830 was nearly cleared of woods and nearly overrun with sheep) is mostly woodlot this century — timber left to grow as it will, barring invasions by loggers. In

New England one can be alone, walk for hours, even days, and never confront the recent works of humans.

Iowa is manicured. One is never away from people there. It is tended fencerow to fencerow, and there is a movement afoot among farmers to get rid of the fencerows in order to accommodate heavy equipment more efficiently.

There is an asepsis about Iowa. It only starts with abolishing fencerows. The mania to manicure extends to architecture and home decor. Nothing is very old in Iowa. What there is of indigenous tradition seems eclectic, recent, and slight. Iowans seem set upon eradicating irregularities of all sorts. After a while, it seems that all walls are washable. All floors are tile or shag rug, all houses, save Joe's, have recently been sheathed with vinyl siding.

"Visitors in Creston are always enthralled. . . . You're in a wide-awake town populated by people on the go," says the Chamber of Commerce brochure.

It is Joe's chore day in town. After we have dropped off the realtor, we head for the bank. Joe has to make a payment — a payment that combines loans on the grain that he is stuffing into the current batch of young pigs, on the new $50,000 combine that stands idle while it rains, payments on the 160-acre piece he does own (he bought it from his mother) and payments on a prefabricated steel machinery barn that he put up a few years back.

In town, Joe knows nearly everyone we pass. He nods to one and all — nothing personal — mutters " 'Lo" and "Fine" as he meets others, and trades quips with a special few. Niceness is the order of the day, week, and year, year in and year out, here. It is part of the code. The banker rises and greets Joe.

"Not warm," says Joe to the banker. They shake.

"Didn't think it was," says the banker. Joe grins. The banker returns the grin, with pretty good timing, waiting a few seconds first and staring Joe in the eye all the while. They discuss refinancing — the next chunk of money buys a new lot of feeder-pigs for Joe to fatten. It is a routine sort of loan — Joe almost always prefers to operate on borrowed capital in order to increase the funds he has available to handle crises of equipment failure and to buy grain or machinery when good deals come along.

Joe operates in the standard manner. Like most American farmers Joe has sharply increased his liabilities as the worth of his land has risen in the past five years. In 1970, farmers' liabilities in the U.S.A. totaled about fifty-nine billion dollars. Five years later, liabilities totaled about ninety-four billion dollars.

Joe hands the banker a check for a couple of thousand dollars. The banker, who went to high school with Joe and watched him play football, shakes Joe's hand again as they part.

"I never came out after you," says the banker now, still smiling.

Joe nods. "No, you never had to," he says.

Both exchange those smiles again.

"Wish I didn't need to do business at the bank," he says as we walk away, touring Main Street. "There's a few around pay for it all themselves but you've got to have your farm all paid up to do it. Credit's the modern way to do business. Use money as a tool, they say, and I guess that's right."

In 1869 there was no Creston, Iowa; in 1870 there was. It didn't spring up by edict of an early visionary (like Salt Lake City), nor by legislative compromise (like Washington, D.C.), nor because a river finally grew too wild for shipping there (like Enfield, Connecticut). It was, from the start, a committee town, and the first committee was expressing a corporate need, translating it into a public inevitability. Surveyors read their orders, stuck flags in new maps, wandered about the rolling prairie with grass up to their navels, stuck a wooden stake in the ground in the place they thought likely to be of greatest commercial advantage — the highest point between the Mississippi and Missouri rivers along the rail route they had chosen — and went home. Someone empowered to do so signed work orders, memos, checks, then claimed, bought, traded and seized ground and said it was O.K. to build on it.

From Creston's first moment, what was good for the railroad was good for the town; the well-being of business was the well-being of her workers, and the workers came to feel staunch about good business. They put their hearts and souls into it, making a place out of noplace, adding the personal history of a thousand lives and a thousand years to the blandness of the corporate need, shoveling up a "heap o' living" on the featureless ground of a boom-town-to-be.

The town was laid out simply in 1869 — first prairie, then a stockyard, a roundhouse, and Main Street. As George Ide put it in his 1908 *History of Union County, Iowa*, when the railroad's work crews reached Creston, "the luxuriant grasses and beautiful flowers of the prairie disappeared."

The story of Union County boosterism rightly begins thirty-five years earlier. The government gave federal land to the railroad and the railroad sold it to settlers — half a million acres went initially at $1.25 apiece, one quarter down and ten years to pay. Some Mormons came for a while, but, unwelcomed, they soon moved on west. A "white" family came and stayed in 1850. The head of household was a blacksmith named Joe Jorman. He made such a good living shoeing the horses of California-bound settlers that he "was compelled to make a Strong Box" for his fees. A parenthetic but telling trace of bias follows next — written at centennial time, 1876, by a C. J. Colby, who seems also to have run the ad-rich boom-town local paper. Colby preferred his founding fathers proper, and counted heathen and immoral births to be part of the long past, not part of the new future. So he noted in his town history, "The August following [following the making of the Strong Box] was born the first white child (except mormon) in the County . . . Charles Lock." A baseball statistic.

They had a rough time for a while, says Colby. They lived on hogs and hominy. A stranger died while passing through, and they had no lumber to make him a coffin. They set right to, though, hewing cottonwood boards for the job. These stories show the early settlers to have been decent, industrious, hospitable, pious, and sectarian.

The next story underscores those traits, and shows the other shore of the same river. Potawatomis, Omahas, Pawnees, Foxes, all lived on Union County prairie — the earliest outsiders lived at their sufferance. One dastardly fellow named Hale-Driggs both thought the Indians of no account and also coveted his neighbor's wife. He murdered a white fellow, then bore earnest witness that a certain Indian had done it. Accounts are muddled, but apparently the accused was "a well-respected Indian." There was a combination mob scene and trial; everyone was worked up. The accused, in a dramatic lament, bared his breast and said he was innocent, that in fact his accuser had done it. The crowd went off to find out what Hale-Driggs thought of this. They discovered he was gone, and the victim's bride had gone with him. They let the accused Indian go. But apparently they remained embarrassed, or regretted their lack of opportunism. As Colby tells it:

A feeling of apprehension obtained a foothold with many of the settlers and the final result was the removal of the tribe to the Indian reservation in Kansas by about 1856.

By 1876 more than 650 persons worked for the railroad at Creston — the monthly payroll came to about $40,000. Main Street grew longer — three banks, ten churches (none Mormon, two Catholic from the start), a steam flour mill, lumber and grain mills, liveries, groceries. At one end of town a jeweler traded in gold and sold diamond rings for weddings. At the other a single ample general store sold plows, perfume, needles and bedding. They also sold coffins and practiced undertaking.

The town grew with the railroad, and the railroad grew with the rising volume of Corn Belt agriculture. More and more

trains carried crops east and goods west. They fixed engines in
Creston, and boarded train crews. Generations of railroad
workers were sired in town, as well as generations of farmers.
Shipping rates, and crop storage rates, access to sidings and
loadings were the hot political issues of the region — issues that
pitted the prosperity of the railroad against the well-being of
the farm families settling in the hills around the town. There
were two sets of boosters in town.

Development of the region's trade was of interest to all. Some
financially minded gentlemen conceived a project in 1888 to
further everyone's interests. They hung a proclamation on
Main Street:

> Recognizing the great benefit the erection of a Blue Grass
> Palace built of the grasses of this section, cured as hay and
> otherwise, would be to Creston and this vicinity, we hereby
> unite in calling a mass meeting. . . .

This was the era when farming went commercial, when
farmers noticed each other and became distinct from suppliers
and shippers and processors and vendors. They allied themselves with the business of selling their crop. Some farmers in
Wisconsin sent the President the world's largest cheese, big as a
cabin.

And Creston built a castle of grass, a hundred feet on a side,
with stocky towers at the corners. A central tower, shaped like
an upside down ice-cream cone, reached up ninety-two feet,
wrapped around the whole way by a circular stair. The overall
effect was spectacular — rather Kremlinesque with horseshoe-
shaped arches, tall parapets, and portholes here and there. As
the county history describes it, "Baled hay was used for gate-
ways and all projections and towers as far as possible, and other
grasses and grains were used in every conceivable way, the
whole effect being a bower of beauty. . . . In size, design,
architecture, finish, and decorations, it surpassed anything

ever attempted in Southwest Iowa." It was gussied up still further before a "Blue Grass Exposition" lent honor to the nation's centennial year. More catwalks, domes, turrets, arches, and minarets were stacked into place, the flags and pennants were festooned all around.

It would be pleasant to say that a wandering herd of buffalo ate it. But the buffalo were gone by then. It burned a few years after the fair. It is still the chief event of the sort of Creston history that is written in books.

Joe Weisshaar's foreparents missed all the fun. When they showed up in town in 1910, the railroad crowd was very much in the ascendant. They'd left Assen, near the Swiss border in Germany, in 1882, and stopped emigrating when they reached Chicago. Members of the family worked there for a few years as housekeepers, carpenters, cooks and coachmen, then most of them set out for the country. They moved south through Illinois and Iowa, "so poor," according to Joe's Uncle Rudy, "that we greased our wagon axles with butter because we couldn't afford grease." They survived "two panics, and always voted Democratic" — a lesson learned during their first days in the German Catholic wards of Chicago.

Joe's grandparents, still mixing English and Schweizerteutsch at home, farmed in a little town called Tainter, Iowa. The family settled in Creston and things got better. "It used to be," Uncle Rudy recalls, "that people would help each other a lot more. Nowadays it's dog eat dog and devil take the hindmost. If you're farming a quarter section, somebody's trying to get the damned thing now. One fellow, I remember back then, the horses got distemper. Other folks did his plowing, and put the crop in too. Once if you were sick, the neighbors took care of you. Now, if you're lucky, insurance pays for it. We lost country life when we switched to tractors — you worked alone more. But we really lost country life when they started closing the country schools. You'd get down below eight students and they'd combine schools."

Creston was in the last throes of its commercial glory when Joe was a boy. It had the world's largest roundhouse, and steam engines were always coming and going. About 1945 a tornado destroyed many of the railroad shops. The coming of the diesel train had already ordained the demise of the company town. Diesels packed so much more power that they pulled longer trains and heavier trains, and could pull them farther between fuelings and waterings, and farther between servicing layovers. Manpower requirements lessened.

During the Second World War, according to the son of one old railroad family in town, railworker-farmer jealousy revived, but for the first time it went the other way. Farmers got draft deferments; laborers went to war. The tension may have been exacerbated by the German roots of the farmers. As Joe's Uncle Rudy said, "I was disgusted when FDR decided to go to war against Germany. Some folks tried to give us a hard time for being of German origin, but we didn't give them any chance to do that. Yes, we put a stop to that."

Today the railroad workers have ceased bothering to be boosters. One old fellow in an engineer's cap said to me down at the station, "When we go, there won't be any to replace us. They don't need us anymore." They also don't need their station anymore. Amtrak's *San Francisco Zephyr* stops in Creston at a surprisingly civilized hour, and fertilizer and grain still move by rail. But today the railway office is one of those glass and metal sheds Amtrak has taken to leaning against the disused hulks of the once-monumental town stations. The railroad recently offered the station to the town. A committee of latter-day boosters, which included the Weisshaars, after a long struggle, succeeded in having Creston accept the gift. The station, a fantastic bit of rococo revival architecture, houses a hall, town offices, and a recreation center for the town's increasingly solitary youngsters.

For the first time in eight long days it's stopped raining. The wind comes up and blows a hulking herd of clouds out of the sky. The sun shines for a few moments every once in a while. Still, harvesting cannot commence until the sodden fields have had a week to dry. Frost has browned the corn. It's ready to take, safe in its husks — what there is of it following the drought. The soybeans sag, but Joe says a combine might still recover forty bushels to the acre. At this point in the wet fall he still harbors some hope — his year's income is still in the field.

While he waits, Joe looks for small chores, repairs to attend to that he has neglected in the rush of the growing season. The job for this afternoon is to fix the collapsing roof of a cattle shed tacked to the side of the oldest barn on Joe's home farm, across town. His mother lives on the farm, alone. Joe says a post supporting the shed roof has rotted.

As any hand worker will attest, an hour of setup time is worth ten of disorganized doings. The more powerful the tools, the more valuable the setup time. Joe has a genius for closing his eyes, considering a job, imagining the tools required, the snags that will befall the worker, and how to work smoothly. He thinks, and lists form in his mind — of parts, tools, personnel, the steps required to do the job.

What he assembles for fixing the cattle shed are (1 and 2) hammers and (3 and 4) crowbars, one short, one long, and (5 and 6) tamping bars, (7) one beam, 6" x 12" x 12', and (8 to 10) assorted 2 x 4s, (11) nails, 10d and spikes both, then, (12) one slow and steady farmhand named Dave, fresh out of the local agricultural junior college and agreeable to every hour of work, at $2.50. Dave drives off on (13) a 90-horsepower Massey-Ferguson tractor. Fastened to the tail of the tractor is (14) a giant earth drill — an auger mounted in a vertical frame that will empty out postholes a foot in diameter and ten feet deep. Dave leads the expedition to the home farm, arriving and parking by the drooping roof. He looks down the road and sees (15) Joe, clattering toward the worksite on (16) a bulldozer, hauling (17) a dumpwagonload of (18) dirt. Joe dumps the dirt near the shed, then sidles the dozer into position, and lifts its blade up under the drip-edge of the roof until the dozer and not the rotten support post carries the weight. Dave takes in hand (19) a small chainsaw, and makes short work of the rotten post, cutting a chunk of it out just above the ground and leaving a long stub dangling from the roof. Together, Dave and Joe lug into place and secure with spikes the beam that will carry the roof. Dave backs the tractor to the edge of the shed, right next to the dozer, lowers the auger, and in a moment has excised a fat, deep plug of earth. The dirt seems to pour from the ground, settling in a small embankment around the whirling auger. Dave trims up (20) an old phone pole with the saw, then Joe and Dave horse it into the hole, heave dirt around it, and tamp the dirt down snugly. Joe leans over the bulldozer and works the controls. The blade drops down. The roof creaks as it eases onto the new post. It doesn't sag. Fifty thousand dollars' worth of equipment. A score of items on the parts list. One hour of time, taken from proper farm chores because the ground's too wet to permit fieldwork. The old shed's good for another twenty years, if the farmer is.

On paper, Joe says, it looks like he could hold an auction tomorrow ("would probably be a good idea, too") and sell everything he owns and gross a quarter of a million dollars. Still, as farms go, Joe's is a low-budget operation — a no-frills sort of place where purchases are made with caution. Joe favors a middle road in implement buying. This policy has left him with a large array of battered but sturdy equipment that he obtained for a cost far less than its replacement value. He shops for slightly used machinery, then babies it. His inventory includes six tractors, a Massey-Ferguson 1150, an M. F. 1130, a Case 930, an International Harvester 606 and loader, an IH 560 and an IH 340 Crawler. Also in his inventroy are the huge Allis Chalmers combine, an IH field chopper, a forage wagon, three other wagons, a grain dryer, two sets of IH plows, two balers, an ammonia applicator, a 15-foot Landall Chisel plow, an implement trailer, a John Deere planter, an IH cultivator, a Case field cultivator, two feed grinders, three augers, two hay elevators, two grain bucket elevators, three bulk feed tanks, five hog shelters, twelve hog feeders, two cattle feeders, a Heston round baler, a '73 Ford 18-foot cattle truck, a '68 Chevrolet dump truck, several pickups, one family car — on paper, $92,450 worth of equipment. Then there's land, buildings, and goodwill.

Why does Joe have so much? Because of the timeliness required for successful farming today, and because the sort of farm that Joe is running involves two separate operations, each with special machinery, one growing feed, and the other dispensing it to fatten animals, which are then sold off.

Joe has assembled a yard full of machinery with very particular properties. Joe, of course, buys what he needs. Practicality is an ingenious allocator of scarce resources. Joe's many and powerful tools symbolize the economic pressure on him to farm *efficiently*, in a special sense of the word. The tools fit the job and the tools complement one another — the grain wagons are the right size to receive a day's feed from storage and offload it at the various feeding stations along the outer wall of the swine house. The tractor that hauls the big round baler about is large enough for the job. The grain truck and wagons hold enough grain so the combine can unload during harvest and then get back to work without waiting for more trucks and wagons to return. The plows — five and six "bottoms" to each — tax but don't exceed the limits of the tractor that draws them, and the disks, spray equipment, cultivators, fertilizing and manure spreading equipment likewise match the job and the available power.

The array of tools assembled by Joe represents — give or take a late-breaking development or two — the best ways to grow food as *efficiently* as the current arts of agricultural engineering allow. But the efficiency Joe achieves is a capitalist's efficiency. He has assembled a business — loans, inputs, know-how, processes, marketing tactics, as well as tools — in a way that maximizes the profitability of farming, given the amount of land he has available, his capital resources, and the nature of the labor he can bring to the operation.

He has done an exquisite job of allocating limited resources of land, labor and capital prettily. The job he has done might bring an aesthetic pleasure to anyone given to regarding fondly delicately functioning balances of life, machinery, principle, and *urgency*. The urgency — the need to meet loan payments when they fall due — moves the show along; it is what makes Joe Weisshaar eat breakfast standing up and then run out on the double to do chores. It is the force majeure of the whole enterprise, the impetus for Joe's drive to insure the survival of farm and family. It's the same force that drives America.

There are, however, other measures of efficiency than the one Joe is constrained to heed, and they may, in some changed future, come to be of more pressing interest to Americans than they are now. If that happens farms will look different.

One such measure is, of course, energy efficiency. In that respect Joe's farm, like those of most of his neighbors, may not be the most efficient. Hunter-gatherers survive in a food supply system to which they add no energy beyond their own labors. Farming with horses is quite energy-efficient. And the pattern followed in some poor nations, of using animal power, local wind and water power, and some small internal-combustion power, has proved adaptable to industrialized situations where there is cheap labor and not much available capital. Researchers estimate that modern American farms spend about half a calorie of energy for every calorie of food they produce. Another six or seven calories go into processing, shipping, packaging, et cetera, of food, making, in aggregate, a net energy loss on food consumed in America. Exponents with an interest in supporting things as they are argue that one can't eat petroleum. But one certainly can eat food bought with money, and as energy costs rise relative to total cost of production, less energy-intensive methods of farming, and of food processing, may well come back. One shift, away from some farming methods that use "embodied energy" (in the form of nitrogen fertilizer, for example) is already underway. Farmers, Joe Weisshaar included, moderate fertilizer use as prices jump. At the same time, they pay more attention to efficient on-farm manure distribution, and to crop rotations that add soil nutrients.

Joe is, quite naturally, interested in growing as much corn, as many beans, as great a carcass weight of hogs as he can because crop income is his bread and butter. But the equipment array and farming methods he brings to the chore are not efficient for maximizing production — they do not represent the way to grow as much food as possible on each acre of his land. This is, of course, not a personal failing of Joe Weisshaar's — he is a good farmer doing what Iowa farmers do to profit enough to pay their bills.

But the fact remains that if Americans come to desire that efficiency of production be emphasized, we will not farm at all the way we farm now. As the formidable agricultural economist Harold Briemyer puts it in his recent *Farm Policy, 13 Essays*, modern agricultural machines

... let one farmer cover vast territory. But he does not cover it very well. Further, mechanized farming is increasingly confined to terrain that accommodates large equipment. It is really a land-wasting technology. If instead of using industrial machinery mounted on rubber tired wheels the United States were to adopt Japanese style farming, cultivating intensively every pocket of productive land, what would happen to total output? The guess offered here is that it would climb significantly.

Joe's farm does not grow as much food as might grow there. At present it would cost him too much to do so.

It is "efficient" in fitting into the complex contemporary pattern of input availability, labor "scarcity" and costliness, availability of technological systems (both stuff and savvy), in tapping into community storage and marketing structures in ways that make Joe a living.

Beyond that, Joe's farming style helps make the banker, the

shippers, and processors their livings. Also the tractor plant workers, salesmen, repairers, and farm laborers their livings. And, in a sense still more removed from Joe's world, his capital-intensive/low-labor farming style makes livings for the ex-farm families who grew food in Iowa back when farms were fifty acres instead of five hundred, when an earlier technology defined a smaller scope for a worker's labor.

The efficiency that Joe's farm demonstrates mirrors the energy-intensive, resource-, capital- and land-rich system in which it exists. It reflects the spotty distribution of wealth. It reflects rural living conditions, and urban ones as well. Joe's methods not only feed city people, they "free" country folk to become city people as "efficiency" drives them out of the countryside. Any mandate to increase either energy efficiency or productivity of farms would require more labor-intensive methods than does the industrialized farming of today. And if such a shift were to happen, it would of necessity be accompanied by a corresponding loss of urban workers. As Briemyer reminds us, nothing is simple about "returning to simplicity."

Joe Weisshaar not only frees rural dwellers to go to work in the city, he also goes there himself once in a while to see how former country dwellers live when they leave the farm. He visited New York on holiday in 1977. "New York amazes me," he says. "I sure couldn't live there, like they do. It's hard to fathom eight million people, most of them doing nothing but serving each other. Parasites — they feed off each other. They could float the whole island away — especially Wall Street — and it wouldn't be missed. There was this poor old guy I remember so well. Imagine this. A guy for forty years sits in a restroom down under a big hotel trying to sell you a towel." Joe knows that modern farming "efficiency" has something to do with the guy pushing towels in the men's room. He blames the fellow's plight, in fact, on "the loss of the family farm."

"I see the direction farming has had to take as a structural problem. To change things for the better, you have to work on the structure. I'm a county organizer for the National Farmer's

Organization — it's like a labor union. It's a way of changing the structure. City problems and corporate power all stem from the loss of the family farm. It enabled things like the Viet Nam war to get a toehold in the first place. It's the breakdown of the family, and it's a planned rate of closing out the old system. They do it in order to keep a labor pool moving in — farmers coming into the city were the finest labor pool ever — trained to work. Now those guys' *kids* are moving into the mindless jobs and they won't stand for it. They'll throw a monkey wrench in the machines to get time to smoke a cigarette, no matter how much you pay them."

At the same time Joe feels it is a "structural problem" that he is discerning, he also feels "it is a conspiracy of the higher-ups. . . . At least to make any social movement work, you've got to have a villain."

Joe's villains are Republicans, and he takes their villainy personally, disputing actively the false contentions of all infidels. "Earl Butz is wrong. I'd like to tell him that farmers work months in the field — there should be enough profit in it to pay for that."

He doesn't like the Farm Bureau, a conservative alternative to the NFO, any better than he likes Republicans.

"Ma's going for a drive tomorrow with the Harrisons," says Mary Jane one lunchtime.

"Who's going to drive?" asks Joe.

"Mrs. Harrison," says the missus.

"She's so old," Joe complains, "and besides that she chews tobacco. And they're both Farm Bureau members."

Joe says that the Farm Bureau — which among other things is the most powerful farm-related lobby in the country — used its ties to the Extension Service to sell insurance, that it sides with industry on issues that affect both farmers and industry, that it only pretends to be an organization to serve farmers.

Then Joe is stricken with remorse. "Don't get me wrong. Take my neighbor Rudy Ehm. Big with the Cattleman's Association. An excellent farmer. Knights of Columbus.

Church. Well mechanized — he's dedicated to agriculture. Farm Bureau."

"It's like saying 'Some of my best friends are Farm Bureau members,'" I say to Joe.

"I get along with all my neighbors," says Joe.

Cooperation among farmers has always been a hard thing for farmers to come by, and the reason is an all-American one — farmers compete most of all with their closest neighbors. There's an old joke I was reminded of on the one trip through Iowa, because I saw it printed on a restaurant placemat: "Farmer: I'm not greedy — all I want is the land that borders on mine." Farmers' movements that have mattered have always been short-lived. They have always occurred when farm prices hit a point so low that the sacrifices necessary to withhold goods from middlepersons and drive up prices do not seem costly, and when prices have a long way to rise before they are high enough to tempt loners to break ranks and fragment the solidarity that makes farmers' revolts work.

But for the first time in history, few enough farmers are now producing the food the nation needs so that it is credible to think of them organizing routinely. Such organizations have started. The NFO is one. Farmers' marketing co-ops are others. There has been a violent shakeout of farmers, and the hicks, hayseeds and most independent were shaken hardest. The farmers who remain, while still not terribly sociable, do demonstrate by their mere continuation in a hard business that they are capable of perceiving mutual advantage. Under these conditions, organization is possible.

"What I like about a person is a hard thing to find and there aren't many people I really like," says Joe. "The ones I like give more than they get from you. Now, you take my friend Morris Smith, for example. He's one of the few.

"He's how I got into the National Farmers Organization in the first place. It was sixty-five. We weren't involved in anything. All we were doing was working. Morris came up to the front door. He almost didn't get in — he was unshaved and he looked like a bum. But he's the most community-minded guy you can find. He's on the board of the local child care center. He's humble about it. Anyway, I listened to what Morris said about selling my animals through the NFO, and I was wary. There was talk around town that you couldn't get out once you were in. You could. There was a procedure — a ten-day period when you could get out every year. They had to do it like that, because they'd signed contracts to supply manufacturers — had to know what was coming in.

"Anyway, I got in on the sixty-six holding action. We didn't market anything for the longest time. Just kept market-weight pigs right here. I had to go out and borrow a lot of money. But we did run up the price. I got to like the NFO. I asked too many questions — and it didn't make me popular always with the management. But it was democratic. Most decisions then were made at the grass-roots level. Two thirds of the members still must ratify important decisions. I hear Farm Bureau decisions are handed down from the top. Every member has a vote at the NFO convention. So I became a county president — to help get things moving around here.

"Farmers won't get their share until they are guaranteed cost of production plus a fair profit. Farm prices always have to be supported to be at a high level. Farmers have to organize to get a good price — to pay attention to their marketing in an organized way. That's what the NFO does. We market our hogs through their 'checkpoint' down in Stringtown. I used to get down in my knees and say 'What'll you pay me?' but not with the NFO."

It's drizzling again. The morale of the household, which waxed as the fields dried, wanes as they dampen. General Joe attempts to rally the morale of the troops at breakfast. Julie, who is sixteen, redheaded, precocious, star of the senior play, dangerously arty for Creston, heads off to school early, then has to rehearse — it's still *Guys and Dolls* here. Late in the

afternoon, her father tells her, she will try to come back and
if there is no one home, it's because folks might be up in the
soybean fields, "walking beans." She is to join the troops there,
search out sunflowers and foxtail, and destroy same. Mary
Jane will shop, bearing two lists — one for household supplies,
the other for farm supplies. She will come back with hot dog
buns and pig medicine, mayonnaise and a crate of welding
rod, orange juice and a new stock watering tank — "Take the
pickup. They'll load you." Young Jeannie, at nine, needs no
rallying. She is always rallied. She doesn't want to go to school.
She wants to work with General Joe. She will go to school,
then work with General Joe.

Al, the boy. Age: eleven. The white hope of the farm will
go to school, but later. He will go to hockey practice in the
evening, after he too has walked beans for a while. Al already
knows this will be his farm, likes the idea, is already a small
man most of the time. His room is special. It is decked with
African jungle decor — tiger stripe bedspread, imitation lion-
head trophies mounted on heraldic shields, model big-game
rifles he has assembled himself from kits. A pith helmet. Jungle
explorer seems a peculiar aspiration for a lad who is destined to
fatten pigs, who already acquiesces to the difficult role of embodying his father's ambition. Al, who wouldn't like the food
in Africa, who seems not even to like his cereal very much
today, whose real trophy shelf bears not hippo heads but
silvered plastic cups for good sportsmanship in the rink and
fat satin ribbons, gold, red, white and blue, for raising fat
cattle well in 4-H club, Al receives his order of the day.

"Feed the pigs, please," says Dad, "before you head out to
school." He means: back a huge trailer with a fancy automated
self-unloading feed wagon, Al, through a tight maze of grain
bins and feed bins, and load up, then feed out, then put everything back shipshape. "When you get done, don't forget to lift
the grain elevator and to drain the corn out. Then haul the
grinder out of the way — so Mom won't back into it when she

leaves." Al, being initiated. Al nods once gruffly. At eleven, he is already paid eighty dollars a week.

"Earns it, too," Joe has said. Joe trusts Al, and tests him, too, assigning tasks that require intuition and judgment.

"When you're done feeding, Al, check and see if the new lot of pigs is settling in." Al is sturdy, slightly blond, with the handsome but not quite formed round features of imminent adolescence. "Rain's supposed to cut out. I guess I'll send Dave up after we sort hogs. Walk beans with you. Take" — Joe's parts list is strange this time — "Take a couple of cookies and a jug of water. Go to the bathroom first." A good general.

We eat eggs fricasseed in the microwave, and march out to sort hogs. The day is close, biting and cold. A damp drizzle finds its way through rain clothing. Joe wears a bandanna, and raises it up so that he looks like a thief about to enter a bank. We enter the hog house and stand for a moment near the doorway, at the head of the hog chute.

The hog house seems run down. "It doesn't owe me anything," says Joe. Pigs have pushed in some of the low block walls, and Joe has shored them back up. Cracks filled with chaff and scraped manure wander through the floors. The appearance of the hog house, won from years of hard experimenting by farmers and researchers, belies the generation of intricate building design it embodies. The hog house on a modern farm is like the camshaft in an internal-combustion engine. Vector, plan, future action, are implicit in its design. Its cast form leads everything happening around it to happen productively. To use it is to progress toward our national goals.

It is long and low. The long cement south wall rises waist-high and stops. The roof is supported by posts that go all the way up. An overhang the size of an awning keeps the weather out of the open top of the sidewall. Five cross-walls divide the interior of the pen into six sections. They are also waist-high, and run nearly across the building, parallel to the end walls.

The hog chute runs the length of the building along the

north sidewall. Six gates lead from the chute to each of the six pens. Each pen houses sixty to eighty pigs. The pigs can march clear up to the south sidewall. Occasionally, one will put its trotters up on the half-wall and peer out. As Joe, along with Dave, the farmhand, enters the building he can see a pig doing just that — having a look out-of-doors and off across the farmyard. It is standing on its hind trotters, peering, and what it sees is Al, mixing breakfast, Al backing the big tractor up to the hammer-mill, leading the auger into a grain bin, loading up, remembering to lift the end to drain the auger, storing the auger, climbing back up onto the tractor, then driving closer and closer along the open side of the hog house. Finally he drives right past the pig whose trotters are on the half-wall. Al stops six times and each time switches on an auger that ushers ground corn, wheat (bought cheap last year), cooked soybean meal and diverse additives from feed wagon to six feed hoppers, one in each of the six sections of the hog house.

Joe figures one flap for every four pigs. The flaps cover sections of a doughnut-shaped trough that rings the feed hopper in each pen. The pigs nuzzle up to the flaps, raise them with snouts, grab a few bites, and let the flaps clang down again. The sound of clanging pig feeder flaps continues night and day — a perpetual Balinese pig orchestra.

There is nostalgia in it. I once returned to Iowa with a native son who had "been away," and as we approached his home farm, he said, "Oh, I'd forgotten about that sound."

And Joe's elder daughter, Julie, at breakfast had, after discussing the senior class play rehearsal, turned her stately attention my way.

"Have you figured out how you're going to start the chapter you're doing about us?" she asked.

"How would you?" I said, ever on duty.

"Let's see. . . . 'As the sun rises slowly on another Iowa morning, I awaken peacefully to the sound of the clanking of the hog feeders . . . and the smell. . . .'"

Midwestern farmers feel it incumbent upon themselves to

mention to guests the smell around hog farms, to assure visitors that they don't like it either. Surprisingly, considering the smell — which, while not strong, is a constant presence — the pigs are clean. The food is clean (the flaps help), and the floor of the pen is clean. The secret of this good housekeeping is locked in the half-walls between the six lots of pigs.

The floor of each pen slopes slightly, from both sidewalls downward toward the center. And the floor at the center is not cement, as is the floor of the rest of the pen, but is built of steel planks, slotted every few inches with open strips wide enough to admit pig manure shoved through by the pressure of thundering hooves. The slots are narrow enough so the thundering hooves themselves do not get caught in the openings.

Underneath the center section of the hog house Joe has constructed a large manure-holding tank. Whenever he can fit it into his schedule he backs a tank fully as big as a house trailer up to the end of the pit, pumps the tank full of manure, then hauls it out to his fields. It costs him about as much to haul and spread the manure, and to construct, purchase and maintain the equipment to do the job "efficiently" as it would to buy chemical fertilizer of equivalent value and then spread that. But even if he bought fertilizer, he would still have to contend with the pig manure — and the pig manure is better for the tilth of the soil than fertilizer.

Why do the pigs cooperate? Hog engineers talk about swine "dunging patterns," by which of course they mean where in their travels pigs relieve themselves. It turns out that pigs like to be clean — in some senses — and unlike cows are aware enough of their bodily functions to stay clear of both their feed and their sleeping places. It also turns out that pigs like *a view* while they eliminate. Unlike chickens and cows, they are not only self-aware beings, they also have well-developed aesthetic sensibilities. The secret of the half-walls between the six lots of pigs is this:

Each cross wall is solid cinder block, save for one run of about five yards, where it crosses over the slotted steel section of the floor. That single section of the wall is constructed of woven wire fencing. Pigs, being what pigs are, come here to enjoy the view. They can see down the length of the hog house. The building works. The pigs have been engineered into doing their own housekeeping.

Until recently, hogs lived out of doors, and sows farrowed their young out of doors. Their manure required no engineering and no handling. It went into the pasture lots where the animals lived. The new system demands an investment of time and capital in manure handling. In return, farmers gain control of which crops receive the manure. More important, a single farmer can keep many animals under one roof now.

What Joe loses by this new system is flexibility. His "confinement" buildings are specialized and the payments on building and manure handling setup continue by the year. It is hard, under such conditions, not to keep one's hog house occupied constantly in order to pay the bills.

Pigs in Joe's hog house do housekeeping in passing. For the most part, they are not happy with their lot. They are just as happy as Joe can afford to make them, and that is, by definition, as unhappy as they will tolerate without becoming less *efficient*. Pigs in the wild are daring, nasty, curious, abrupt, opportunistic foragers. In the days of pasture hog operations, the conditions of the jungle hung on a bit. Every old farmer in Iowa has "hog knees" from trying to corner his sows out in the pasture.

Old hangers-on — a goodly number linger — still keep their hogs on pasture. It's a reflection of a rather sophisticated management method called "clean ground system" with disinfected shelters. Young stock live lives isolated from older stock. It's a decentralized system, land-intensive and costly in both labor and fencing.

Nowadays progressive pig farmers keep their animals indoors. The roofs, slatted floor manure-handling facilities, cement, plumbing that Joe requires are all expensive to purchase, to put in place, and to maintain under the rough handling of hogs. The cost of "the facilities," as Joe calls them, is assigned to the hogs, whose fate it is to pass brief lives there, and since plant costs are a considerable part of the total cost of hog production, farmers are virtually forced to pack into their hog houses as many animals as can be crowded together.

The limiting factor is stress. The luxuriously authoritative, very state-of-the-art text, Swine Production in Temperate and Tropical Environments, by W. Pond and J. Maner, recommends from three square feet of living space (for pigs fresh away from the mother) on up to eight square feet per animal (for two-hundred-pounders). When one figures that a market-weight pig is indeed about four feet long and two feet wide, it becomes clear that pigs are most compliant to the needs of farmers. A pen of pigs on Joe's farm is like one of those Chinese number puzzles that hold fifteen numbered squares in a rack big enough for sixteen numbers, and by sliding numbers into the empty slot, one is supposed to arrange things in order.

Whichever pig is farthest from the feeder is the number farthest out of place. Hunger is the manipulative thumb and it is an active principle. The pigs at the feeder are like stuck squares — they are hard to move away. With four pigs allocated to every feeder flap, the puzzle is never solved, the pigs are always in motion, in combat, and in a state of stress animal scientists would term "tolerable."

Animal scientists introduce stress to pigs' worlds and consider the results. They have kept pigs in the dark (onset of puberty advances a bit) and in constant light (nothing new), in jungle dampness (rate of gain slackens over 85 percent humidity — great tolerance before that), noisy conditions (loud constant noise O.K., loud sudden noise excites), have kept pigs in sweet and noxious-smelling quarters (they don't care until the atmosphere gets poisonous), have stored pigs in hot rooms and cold rooms ("zone of thermoneutrality" rather broad for older animals

— newborns fussiest, weanlings fussy — but after certain point
heat devastates pigs because they can't sweat).

They have packed pigs in together closer than pushers pack
rush-hour riders on the Tokyo underground, and that is the
most interesting stress of all to modern hog growers, because
pen space comes dear. Pigs will sooner bear stress than take
things out on each other. Once pushed to extremes, their response to crowding seems to be a signal appropriate to the
competition of open feeding grounds. They bite each other's
backs and they bite each other's tails — clean off if they can
manage it.

Then the tail stumps infect. The afflicted go off feed. Rate
of gain slackens. Time is money in the pig shed, as elsewhere
in America. Stress slows. The maximum bearable crowding of
the fastest-gaining hogs available makes a farmer his living.
Just a touch of backbiting, like a touch of mastitis in the dairy
barn, signals that all sail is spread, that no one, in any other
barn, can be doing any better, and that's what's important.

In Joe's hog pens, one pig or another always looks like it
has been beaten with barbed wire. It is always one near the
bottom of the backbiting pecking order. It's easy to spot the
poor creature, careening above the crowd like a number displaced from the formalized head knocking that accompanies
progress toward the feeder in the Chinese puzzle of eternal
hog motion.

"If it gets bad enough, it's move the victim or lose him,"
says Joe.

For all pigs' contrariness, it is neither coincidence nor historical quirk that farmers raise pigs. Pigs are better suited to
the job of becoming meat than are other animals (save
chickens). It's not mere happenstance that we don't eat guinea
pig roasts, raised in a vast farming area where corn–guinea pig
farming is the most profitable for the terrain and climate. Nor
is the reason we eat pig meat simply that we are used to it,

that our grandparents liked it too; if sloths fattened better than pigs in Iowa, we'd have sloth and eggs in the morning and Grandpa would have done the same. Pigs (and chickens) put on more weight per pound of feed consumed, than do other creatures.

This leads smack into an unhandily basic question — a question about why things are the way they are. Farmers, unwittingly heralding the genius of the capitalistic system, flagged hogs as good gainers. The impudent question that follows, about the very order of things, is: why are hogs such good gainers? Or perhaps it's, how did *hogs* get to be the good gainers they are? A second half of an answer, reflecting recent history, is simple enough. Farmers have selectively bred hogs for several thousand years in order to improve the animals' feed:gain ratio.

The first half of the answer, dealing with predilection to gain efficiently, is the more difficult: why were hogs what got bred up? This question turns out to be on the order of why leaves are green, and why it rained today. It's one of those scientific questions about a system so complex that while an answer exists, it is general and speculative.

An evolutionary biologist at the University of Pennsylvania, Daniel Janzen, has come up with a strange and delightful tale that makes a pretty sense out of dark millennia past. As far as I can understand (from an article called "Why Bamboos Wait So Long to Flower," which was in the 1976 *Annual Review of Ecology and Systematics*), Janzen's argument on the subject goes like this.

Pigs and, of all things, bamboo coevolved in the same Indian-Asian region in a way that was crucial to pigs' nature. Bamboo, it seems, is a peculiar plant. It reproduces itself by rhizome, asexually, developing "cohorts" that sometimes cover many square miles, all of plants whose chromosomes are identical. But at occasional intervals, it also reproduces sexually and thus maintains genetic variability. What lives in nature has competitors and predators, and evolves responses to competition and predation. Bamboo is tough, grows in dense stands that smother other plants, et cetera. Its style of sexual reproduction also turns out to be a defensive tactic.

Bamboo is a "mast seeding" plant. It waits for many years. And then, suddenly, an entire vast cohort of bamboo flowers sets and drops seed, and dies, in a season. The synchronicity of bamboo cohorts is astonishing. Cohorts may wait half a dozen, twenty, fifty, even a hundred twenty years between seedings, and then seed in concert. Cuttings removed to different climates on different continents maintain the home rhythm, so the clock is apparently internal and not set by some environmental cue.

Janzen asks, "What is the adaptive significance of synchronized seeding by bamboos?" The answer is that it represents an unusual but successful tactic for dealing with the predator problem at seeding time. Some kinds of plants develop poisoned leaves, some get to tasting bad, some grow stickers, some encase seeds in indigestible packages and encase those packages in delicious packages; to eat them is to spread the plant's gospel, and add fertilizer to it in the spreading. Some plants time innocuous seedings so that something tastier seeds at the same time. Some seed so early or so late that predators haven't arrived, or have gone away.

Bamboo is a plant with which many of us can identify; it employs the tactics of a bountiful earth-mother. I had a grandmother like that — she spent a great deal of time preparing food (each meal was a banquet) and then served a feast of such proportions that even when the neighbors were called in, and even when they called in their own neighbors, there was still plenty left over.

Bamboo, like a botanical grandmother, moving to a slower clock, cooks for years and years without serving a meal, or at least, spends many years conserving energy by not making seeds, but by growing strong and fat instead. Then, all that stored energy is suddenly converted into a feast of nutritious seed and is dropped.

The reason why bamboo waits so long between feasts is precisely to make sure there is too much food on the table. When it finally seeds, bamboo issues hundreds of pounds of edible seeds per acre — a very respectable grain yield. Were such a feast available every year, the predator population would rise until it was able to consume all the seed bamboo might give out. But because the mast seedings are occasional, they do not contribute to determining the level of the predator population. The seedings always exceed its needs. Bamboo's tactic is to offer "predator satiation" at the existing population level, and still to have some seed left over to sprout. What this translates to is a vision of thousands of jungle creatures snoozing in a dying bamboo grove, their potbellies swollen with bamboo nuts until they can scarcely crawl forward to have another of the delicious morsels that still lie about the cohort.

In Janzen's refined picture, bamboo seeds for a whole season. The predators arrive, can't consume the midseason glut, but do consume the tail-end production of the plant. This serves to reinforce the synchronicity of the next generation of bamboo.

Quirks of Asian history can apparently be explained by reference to stupendous bamboo mast croppings, which have drawn animals from hunting grounds where tribesmen usually found them, causing massive human starvation to accompany the massive animal gluttony.

Janzen next fancies the qualities that might be found in an animal successful at taking advantage of these occasions for feasting. It would be a highly mobile animal, so it could travel between asynchronous cohorts some distance apart. It would develop a great capacity for going without and for storing fat, because going without is the usual fate of those waiting for bamboo's feasts. It would be omnivorous, so it could eat as well as possible and tide over in the times between. It would have a relatively low sense of territoriality, so its kin and others could feed simultaneously nearby. It would gain weight extremely quickly, so that whenever occasional periods of feasting happened, it could make the most of the festivities.

It would, furthermore, use the occasions of feasting to reproduce, a tactic that would, in a manner of speaking, invite the family to join the feast. It would produce many youngsters in each opportunistic litter. And the youngsters would have a short gestation, then be early to wean and join the chow line.

In the tropical woods of India and Asia, the pig was born and evolved. And overhead flew the jungle fowl, grand-ancestor of the chicken. There is occasionally free lunch in Mother Nature's world. Farmers have taken over the duties of bamboo plants. There is no free lunch in the world of American commercial agriculture.

As Al stops at each of the six pens the pigs strain forward. They move for the entertainment of viewing him, and not for any promise of food. There's always food. In Iowa the bamboo has always just dropped seed; hogs eat as much as they desire. What they eat is controlled with the same efficient eye for science and economics that governs the efficiency of the farm as a whole.

Like dairy cows, pigs eat for a living. Navigating in the crowded world of total confinement, hogdom is their chief activity. Joe has hung a rubber tire on a chain above each of the pens. It dangles at snout level, and occasionally a bored pig will try to scratch her back on it. It will move, and the pig will shoulder its way over to the pen dividers instead.

"That's why they're all cracked up," says Dave, the farmhand.

Dave trudges down the narrow chute that runs along the side of the pigpens. He studies the animals in the last pen, counting with stubby index finger pointed at pigs, mouthing numbers to himself. He moves down one pen and counts more.

"We'll get 'em all from the last pen — few in number five ready too," he says to Joe.

The hog house whispers with the sound of pig bristles scraping on pig bristles. At any one moment at least half of the four hundred or so pigs in the hog house must be brushing past the other half, in transit between scratching posts, feed flaps, toilet, and sleeping platform.

We walk outside, lay hands on a structure that looks a full-scale model of a colonial era aircraft boarding stairway hewed out of rough-cut lumber and pinned together with spikes. It is a Jacob's ladder for pigs, and pigs are most particular about where they want to go. The loading ramp has solid sidewalls, because pigs don't take to heights. Its incline is rather steep, but is firmly braced, because pigs abhor floors that bounce. We wheel the ramp up to the chute that extends from the hog house, then Dave backs a big stock transport truck up to the high end of the ramp. The wire dangling from the tailgate turns out to be there for the purpose of linking ramp to truck when Joe and Dave repeat this procedure every two weeks.

Jeanne, Joe's youngest, the family hellion, gallops up on a large white horse. She shouts, "Whoa, Winkie," dismounts, ties Winkie to a mowing machine that lies half-disassembled near the hog chute. She climbs up on the roof of the truck cab and perches down to watch the show. With her have also arrived the guardians of the farm, two lean Doberman pinschers named Tina and Cici. Jeannie shouts their names to them. They bark and prance. Winkie eats weeds. Jeannie spies Dave, who is an amiable being, pneumatic, tubular, like the Michelin man.

"Hey, stupid," she shouts to him. Dave looks up, shakes his head, says, "Stupid? That's you."

Jeannie squeals.

"Dave," she shouts, "you want that scraper in your face? Pigs, watch out. Eat. This will be your last chance to eat."

Dave and Joe, mask raised, reenter the hog house, Joe carrying a section of "hog panel" — a length of slatted metal fence like the one affording the toilet area a view. Joe swings open the gate of the last pen. The gate turns out to be just the right size to block off the east end of the chute that runs the length of the building, leaving pigs a clear path only to the west — toward the loading ramp.

The men enter the pen. The hog panel is also the right length. It is one pig-width shy of the total width of the pen. I enjoy these moments of puzzlist's delight, of seeing suddenly the simple design of the intricate modern matrix that manufactures pigs.

Hog sorting is a reflex sport, like heavy-duty Ping-Pong.

The truck wants thirty-two pigs — a thirty-third would have to ride on the back of the others. It is the right-sized truck for Joe's operation. Joe has contracted to sell his hogs to the NFO collection point twenty miles away at Stringtown. The NFO has contracted to sell this or that packer uniform lots of hogs. Optimal weight is about two hundred thirty pounds, because at that point pigs butcher into nice familiar-sized chops, bacons that fit into the supermarket meat cases without hanging out on the floor. Pigs of that weight are adolescents — about five and a half or six months old. Left to mature, they will keep gaining until they reach about five hundred pounds. The meat of a five-hundred-pound hog tastes fine, but it tends to be fatty. In the old days, hogs used to be overwintered and brought up to mature weight. Nowadays consumers want meat and not lard. It turns out that once past the magical weight of two hundred and thirty, the ratio of fat to muscle in each additional pound of gain begins to increase. Modern hogs continue to gain about a pound of body weight for each three to four pounds of feed (nearly twice the gain per pound of feed hogs got a century ago), but the ratio of pounds of feed per pound of lean carcass falls off. Two-thirty is the magical weight. The assembly point docks farmers a few cents a pound for animals exceeding that weight. The rule motivates Joe to strive for the uniformity modern pork packers demand.

The pens contain progressively heavier animals. The last pen holds animals reaching market weight. As Joe advances, he pushes the hog panel before him. The pigs retreat toward the toilet area, climbing up onto the feeder and sliding off, mounting each other as the press toward the rear wall continues. Joe is alert. He watches the flow of the herd. Its collective irritability

overpowers its collective timidity as Joe presses it still harder.
Hog sorting emphasizes the precontact moment. It reminds
one of wrestlers stalking each other, feinting, drawing back,
moving in quickly when conditions are right. The nearest hogs
spot the pig-width opening to the side of the hog panel Joe
carries. The hog that starts through it appears still underweight
to Joe, and he shoves the panel a few feet to the right, closing
off its way out. This creates a new opening to Joe's left, and
another hog heads for it. Joe doesn't like the looks of that one
either, and he hefts the panel back to the left. By this time,
the pigs on the right flank have ridden and jostled each other
into a new arrangement — they move around fast, like a school
of trapped fish patrolling the walls of a tightening seine net. A
blatant fatty is now nearest the opening. He charges and Joe
lets him pass through. Dave slaps his butt as he runs by, and he
trots right on, hesitates when he comes to the accustomed limit
of his hog world — the road to the aisle. Dave slaps him again.
He goes out. Joe has let another pig through. It follows the
first pig. The free pigs slack off as they leave the excitement
behind. They wander down the chute slowly toward the open
door and the loading ramp, sniffing at the pens they pass.

As more and more pigs stream out into the chute, Dave
moves out after them. There is more room for Joe to maneuver
in the pen, and his moves become more refined. He goads a pig
with body English, forcing the one he wants toward the leading edge of the pig mass, then lets it through. A skinny hog
slips by and Dave heads it back in. "This little piggy goes to
market, this little piggy stays home." A few more candidates
are allowed past, and Joe is done.

Joe presses the hogs in the aisle out toward the ramp.
Meanwhile Al has finished feeding. He has parked the tractor.
He has hauled the grinder out of the way — so Mary Jane won't
back into it. Mary Jane backs up a pickup truck, on her way to
the vet's now with a pair of hundred-pound boars that should
have been castrated at birth but somehow managed to slip into
a lot of pigs with their sex intact.

Al approaches the loading ramp. The lead pigs mill in its narrow opening, peering upward toward the truck, shuffling away as if that weren't really the only place left to go. As the chute crowds with market-weight hogs, the alternative begins to appear increasingly attractive. Jeannie watches attentively from the truck cab roof. Al takes in hand a cattle prod. It looks like a small bicycle tire pump.

He climbs over the wall and into the chute. He seems unexcited, professional, has obviously done it a hundred times. The pigs are less experienced. They are agitated. Al waits until the lead pig is gazing up the loading ramp, then touches her ham with the live end of the prod. She screeches, long, breathless, astonished, fearful roars, screams in the piercing registers. She sounds like a soprano lion. It works. She clambers up the airline steps, still roaring. With the same urging, the next in line follows.

Dave keeps counting pigs, pointing and counting. Finally he shouts above the screeches and grunts, "O.K., that's thirty-two of them," to Al and Joe, and the hog sorting is over.

The men slide the tailgate back into place, roll the loadiing stairs away, and then go in to tidy up. Al goes to school.

The truck stands alone in the yard, low on its springs, bearing a shuffling cargo of fat hogs that still grunt and squeal. After a while, Dave comes out of the house and drives off to market.

What he provides with transportation is one lot, four tons, U.S. #1 grade hogs with a sprinkling of U.S. #2s thrown in: "Swine near the borderline between the U.S. No. 1 and U.S. No. 2 grades are thickly muscled in the hams, loins, and shoulders. The width through the ham is nearly equal to the width through the shoulders and both are wider than the back. . . ." (from *Official United States Standards for Grades of Slaughter Barrows and Gilts*).

The name of the struggle of hog farming is carcass yield. The term refers to the amount and sorts of usable flesh on a pig.

The demands for parts of pig has changed over the years, and the nature of breeds has changed to keep pace. In 1886, when Mom still made apple pie, and when she was still sure it ought to be made with lard crust, hog authority James Long lamented, in *The Book of the Pig*, that "the present system of breeding has materially conduced, not only to the larger production of fat . . . but to the reduction of lean, and this must in measure be considered as a drawback to the value of the pig as a meat-producing animal." This was a soothsayer's remark, written in an age when fat pigs still lurked under every stable, stood caged in workers' yards, and when smokehouses abounded and farmers raised, slaughtered, butchered, processed as well as vended their own products, and treasured salt pork and lard.

We live at the hind end of a short age of fat. Just since 1964, for example, the United States' per capita consumption of lard has halved — it's now down to about three pounds a year. Nowadays portable electronic fat-sonar zaps every pig in the marketplace to determine thickness of back fat.

In the high-stakes world of raising swine stock to "improve the breed," the boars whose issue grow up with long, lean hams and chops with broad solid eyes (not to mention boars whose daughters throw enormous litters, mother well, breed back easily, and whose issue gain weight rapidly) are the hogs that win the race to supply high-price hog genes to the future pig population of the nation.

An air of hucksterism permeates the modern world of hog breeding, and concerning fat, boosters of the Hampshire breed proclaim, "It is hard to find one of our boars with over one inch of backfat," the Duroc boosters say its "pigs are low in backfat and convert feed to lean meat very efficiently," and Landrace boosters boast pigs that are "well muscled."

The antifat turnaround of the hog market was underway by the 1890s. It had to do, first of all, with the emergence of commercial refrigeration. The attraction at the butcher shop (according to Charles Towne and Edward Wentworth in *Pigs from Cave to Corn Belt)* shifted from barreled, salted pork to cooled loins, bacons and hams. The old-style highly bred hog was shaped like a cross between a dachshund and a baked potato — it wobbled about, lumpy, bulbous, and slung so low the belly dragged on the ground.

The earliest efforts to change selected simply for size, and state fairs for a while exhibited half-ton sows and ever larger boars. Their meat was coarse and stringy. More refined breeders sought, instead, small pigs with big hams.

The reorientation of hog genetics in some ways proved to be a political phenomenon. It required discipline of a large and disparate population of hog raisers. The discipline was applied at the marketplace, at the conscious insistence of organizations with economic clout — meat packers and breed associations. They affected the grading structure within the marketplace. They affected the judging standards at fairs and breed shows. They affected the goals recommended, via the U.S. Extension Service, by agricultural academics. And as demand shifted, they affected the choice of genetic stock that common farmers bought to improve their own herds.

S. M. Shepard's milestone work, *The Hog in America*, published in 1886 at the inception of modern hog breeding, brings a bit of sectarian patriotism to the gathering forces for selective hog breeding: "The American breeds . . . lead all the English. . . . They may not be as old, nor be able to trace their family name, '*back to the flood,*' but in a country where there is little veneration for the '*has beens,*' the '*izzers*' are the things sought."

Shepard counseled hog raisers zestily, "Always get the best. Get them cheap as you can, but get them. . . . Don't put a ring in his nose; let him root; the lot will not be handsome, but the boar will. . . . With a hog, the old adage, 'less haste and more speed,' is peculiarly applicable."

The Hog in America quotes "an early agricultural paper" on the subject of the shoddy genetic material farmers had at hand:

I'd just as soon undertake to fatten a salt codfish. He's one of the racers, and they're as holler as hogsheads; you can fill 'em up to their noses, ef you're a mind to spend your corn, and they will caper it all off their bones in twenty-four hours. I believe if they were tied neck and heels an' stuffed, they'd wiggle thin betwixt feedin' times. Why Orvin raised nine on 'em, and every darned critter's as poor as Job's turkey to-day; they aint no good. I'd as lieves ha' had nine chestnut rails, and a little lieveser, cause they don't eat nothin'.

The recent genetic improvement of hogs has been quite startling. When Joe Weisshaar graduated from Iowa State in 1960, in farm management, his texts on hogs told him that he was dealing with remarkable creatures — they gained a pound of flesh from only five pounds of grain. Nowadays, Joe's hogs, better-than-average animals, gain a pound for every three pounds they consume. Joe's college text boasted of the prolific mothers of the swine world, and told him a good farmer marketed five piglets from every litter, and said a good mother would average 1.7 litters every year — at least eight little pigs from every mother each year. Nowadays, Joe's sows average nearly eight marketed pigs from each litter, and bear better than two litters a year.

That makes the hog quite a different creature from the cow, which, after a year and a half of life, followed by a ten-month pregnancy, yields up one delicate calf. Yearling sows already have bidden their first eight offspring adieu.

While Dave drives the truckload of hogs to market, Joe commences the next chore of the early morning — to consolidate decimated hog lots on neutral ground. First he runs the still-too-thin survivors of pen six out of the hog house. Then he rearranges gates, opening new routes for pig propinquity. He drives a few fat pigs from pen five, which was diminished by a sorting two weeks ago, into pen four, which stands empty and disinfected. Then he invites the pigs formerly of pen six to join the former pen fivers in pen four. All pigs sniff. Then all pigs scrap with one another. Three against three in the far corner by the feeder. Two against one in the center. Four against one nearby, all screaming and biting. And eating, grabbing bites even as they tussle. Farmers combining lots of pigs sometimes spray them all with strong scent to disguise the odors that hogs seem to use to mark their own. The combining of lots on territory used by neither previously also helps. Nothing works completely. The fighting takes hours to subside. Tail stubs are worried, backs mouthed and scored. It must be hell to be a modern pig.

There is one last hog chore for the early morning. Joe brings up the manure tank — it dwarfs the tractor. It's heavy equipment. Joe backs the tank up to the manure port, next to the door of the hog house, and he does it casually and perfectly.

He turns on a pump — a very fancy pump that cost a thousand dollars, and from deep under the hog house manure begins to flow. From birth to death, a hog produces about ten or fifteen dollars' worth of fertilizer. It costs nearly that to clean up after a pig, even in Joe's new-style building. The original impetus for hog confinement was indeed to save money in manure and feed handling by installing the pigs in the midst of an architecture designed to minimize human labor.

Farmers then found they enjoyed other savings as well. Confined herds generally have higher "farrowing averages" — litters survive better indoors than out. Disease control is better because, unlike a pasture, a hog pen is small enough to clean and disinfect between lots.

Joe has only recently stopped farrowing pigs himself. "Sinuses," he says by way of explanation, although it is more complicated than that. He does seem always to be using antihistamine nose sprays, and spends a good part of his time heroically ignoring his own discomfort. "Hog dust's what's done it. Moldy grain, feed dust when you grind feed, dried pig crap when you clean up, hay chaff. Farming's no place to have sinus trouble."

When Joe raised pigs "from start to finish" he kept a herd of pig mothers, a few swell dads, and devoted himself both to "farrowing" the largest and healthiest litters he could produce, and to raising the good litters up to market weight. Joe kept eighty sows and marketed about 1,200 finished pigs — over a quarter of a million pounds of pig — each year from 1960 to 1976. Now he concentrates on fattening bought piglets. He buys piglets at forty or fifty pounds from farmers who specialize in raising them.

The new sort of hog business suits Joe's resources quite well in the changing climate of modern farming. It is a transformation in the animal rearing business that many Corn Belt farmers have made. Most of the labor involved in raising pigs is spent in their earlier months of life. Joe's young-stock suppliers do nothing but breed sows, watch over their births and nursing and weaning, then market their children and start again. Farmers taking to this specialized enterprise are usually short of feed. They are farmers with little land and limited capital, but with the finesse and time to devote to the labor-intensive end of hog farming. Feed requirements are low. About three quarters of feed consumption and manure production occurs after the "feeder pig stage," according to a recent Purdue Extension Service *Pork Industry Handbook.*

Purdue estimates the labor required by Joe's supplier to be an hour and twenty minutes for each pig raised to fifty pounds. The labor is spent breeding, attending births, administering postbirth iron shots, castrating males, clipping needle teeth, tail docking, feeding, and, finally, moving the lots of pigs out of their pens, cleaning, disinfecting, and selling. During the entire four months of fattening it will take Joe only an hour per animal to add another hundred eighty pounds of weight to each feeder pig. With his newly freed time, Joe makes more intensive use of his costly hog house, and spends more hours each day working more land with his expensive array of crop growing and harvesting equipment.

Because of the high labor component, Joe's feeder pigs cost him about twice as much per pound as his market hogs will bring. Nearly 40 percent of Joe's production expense is the expense of purchasing animals. That puts him in a capital-intensive fast-turnover business — one in which cash flows more rapidly than in "start to finish" hog rearing. He runs about ten bushels of home-grown corn through each pig, and likes to say that he "markets his corn as pork." Joe has increased his yearly marketing by about four hundred animals since the change. The implications of this change are considerable, and affect many farmers.

Joe has not only increased his marketing volume, he has also increased the vulnerability of his business to fluctuations in the marketplace because he now operates on more borrowed capital, moves more animals each time he markets, and has more at risk in each transaction. His costs for feeder pigs have substantially increased; he now pays for another farmer's labor and risk and profit, as well as for the transportation of the animals he buys.

He has decreased his control of the quality of what he raises, because he has to start each batch with what feeders are available. "You should have seen the beautiful lots of pigs I used to bring to market," he says. He also has increased his likelihood of bringing a contagious disease into his herd. With Joe's herd "open" to outside animals, there is some risk of infection.

Joe did not mean to find himself in this exposed position. He had in mind another alternative to home farrowing that would have served him better had it worked out. The sudden increase in the cost and availability of capital, however, has caused his plans to go awry, to date. Joe's alternative to on-farm farrowing follows a small but important change in the nature of pig raising — one with uncertain consequences for consumers.

Joe, like his wife, is an organizer by nature. When he decided years ago that he liked the National Farmer's Organization, he didn't just join and consent to their dramatic holding actions. He helped organize holding actions. He and Mary Jane decided early that they liked Jimmy Carter, and both of them went on the stump for him. They ended up at the Inaugural Ball in Washington. When Joe decided he wanted good feeder pigs but didn't want to raise them, he started organizing again.

He had heard about a way of organizing farmers to own, collectively, a modern total-confinement, highly mechanized, professionally staffed farrowing house. It's not an obvious thing to do — it takes around three-quarters of a million dollars to bring it off, and that takes costly credit, which has been the problem to date. It takes an understanding of the role of state-of-the-art technology to decide that it's a good business idea. It takes audacity to place so much capital in such a complex and inflexible system. Joe is encouraged by tax laws making it more advantageous to organize on a fairly large scale than to improve things on the home farm. The tax advantage was the deciding factor. As agricultural researcher Marty Strange of the Center for Rural Affairs in Walthill, Nebraska, explains in *Who's Going to Sit Up with the Corporate Sow*,

"Subchapter S" refers to the Internal Revenue Code, and it is an attractive type of corporation primarily because of its tax advantages. The subchapter S corporation may be taxed substantially like a partnership rather than a corporation; that is, profits or losses, as well as depreciation benefits, are allowable directly to the shareholder, and not to the corporation itself. It therefore allows accounting losses to be deducted directly from the investor's taxable income. . . . The investor is also able to defer or delay paying taxes by deducting feed, medical and other costs. These "accounting" losses are taxed later when they show up as profits in pigs sold, but he can continue to defer these taxes year after year as he incurs expenses, until one year he happens to be in a lower tax bracket . . . one reputable agribusiness journal estimates that from one-third to one-half of the initial investment in a confinement operation can be returned in tax benefits in the first year.

The implications of this sort of possibility upon the future of family farming in the Corn Belt are hard to realize. It may seem at first reading merely that ten enterprising family farmers are assuring themselves a profitable source of healthy hogs, taking advantage of the best hardware and the best genes, and have like other businessmen found that they can use a tax structure as it was intended — to promote business activity that is of public benefit. And, in the short run, this is indeed what is happening.

What Joe is proposing to do, however, is part of a pattern of activities whose cumulative effect shifts control of an important aspect of his production — its very lifeblood, the young stock — to "off-farm facilities" and places management in salaried hands. It is only coincidence that the well-to-do gentlemen Joe has enlisted to join him in this venture are farmers who can use the piglets the sow factory will turn out. There are more rich city folk than rich farmers, and there is no impediment to their becoming shareholders as easily as might Joe himself. The skill of raising the pigs removes from the farmers involved to the hands of a hired manager, trained in ag school. The correct practices for him to follow are implicit right in the sophisticated design and functioning of the complex of agricultural management equipment called a farrowing house.

Subchapter S is of greatest use to those in the highest income brackets, and to such high earners the investment may be attractive indeed. Without any shift in "farm ownership" and without the transfer, by deed or even by letting of a mortgage, of one acre of farmland, a crucial aspect of agricultural production shifts out of the hands of ten farmers.

This does not happen because these high-capital methods of hog rearing use capital more efficiently. To the contrary, the results of research at Extension experiment stations show that modernization on the home farm profits an operator more than does a state-of-the-art farrowing house. But tax structures invite the transformation of ownership of this crucial argicultural resource.

It may seem a mere historical detail — who owns the deed to the hog house in 1979 — but this detail in fact is a fit emblem for a remarkable and much larger transformation — in the political organization of America's food supply. This transformation leaves all us eaters closer than we were before to being in the soup.

When economists cast about our economic structure hunting for a remaining "pure competition" sector of the economy, they look at agriculture. Hog farming, for example, clearly fits the four-part definition of a "pure competition" enterprise — one in which (a) producers in the market turn out an undifferentiated product (one farmer's market hog is like another); (b) a large number of producers make the same product (there are about 680,000 hog farms in America); (c) individual buyers and sellers deal in relatively small proportions of the total market (no one farmer sold, and no one institution bought, a very great percentage of the $7.5 billion worth of hogs marketed in the U.S.A. in 1976); (d) there are no artificial barriers to entering or leaving the market (growers and purchasers can produce hogs and buy hogs at will — no controlled production, no rationing, low "start-up" expenses). As a result of these market conditions, until recently the price to consumers of pork has reflected the cost of growing it, processing it and selling it. "Pure competition" has meant relatively cheap food in the United States.

But many conditions having to do with modern technology, modern-sized capital accumulations and modern business structures all combine to make it look extremely likely that in the future, unless an as yet unmanifested "will of the people" works hard at demanding otherwise, food resources will increasingly fall into the control of powers not as subject to the democracy of "pure competition" as those we have been used to in the past.

What is interesting about Joe's plan for a modern farrowing house is the "down-home" nature of the change. It is not a conspiracy. In fact, the chief perpetrator is a staunch family farmer, who previously organized political activity contrary to the sort of change the farrowing house represents.

Asked about this seeming contradiction between professed values and ambition to succeed in farming, Joe offers an honest defense that serves only to show how complex a creature is "rational economic man." He says, as might be expected, that he has "a family to feed," which is patently true, and only remarkable because it dramatizes his feeling of helplessness, his sense of the inevitability of doing what he plans to do, his perception that the alternative to going along with the industrialization of hog farming is leaving the trade altogether.

Once Joe starts to thinking about his plan, he gets even more militant about it. "I should have turned to this idea a few years ago, when costs were less. Should have invested in the futures market back then, too. But you know why I didn't? It was the gloom-and-doom atmosphere of the National Farmer's Organization that kept me out of it. I feel the long-range solutions involve farmers getting together to demand 'cost of production plus a fair profit,' like the NFO says. And I'm a staunch member. But it took me a long time to realize that in the short run, I had to take care of myself at the same time — had to do things in my own business to help my personal situation. That's why I got involved in the futures market, and that's why I got involved with putting together this farrowing house deal. Only trouble is, it's too expensive for local credit people to deal with. I'm not too worried. We'll get what we want."

There is a model for what may happen to the pig business, because it has already happened in another sector of the agricultural economy — the business of supplying the eating public with "broiler" chickens.

The comparison of pig growing and chicken growing is inviting because both are susceptible to high degrees of mechanization. Chickens have lent themselves to "machine husbandry" even more easily than pigs. And chicken growing has passed more rapidly out of the hands of family farmers. In the broiler business the entire chain of production, from feed to bird to processing, to selling, has lent itself to "vertical integration" — to control by a few large investors with the capital and market power to undersell family farmers raising chickens. The result has been the "democratization of chicken meat" at the consumer level. But the result for chicken farmers was just the reverse — chicken production is the sector of the meat economy in which the "economies of scale" of production have shifted most dramatically with the coming of new technology and business organization. A few large processors have come to dominate the industry. "Pure competition" has disappeared. Recently, "product differentiation" has succeeded. Purportedly premium-quality birds, marketed for premium prices, now lie in supermarket display cases side by side wtih sodden, blue and yellow, scrawny, Brand X fryers. The growers of premium birds now control 40 percent of the market — at premium prices. Even the best supermarket chickens seem poor replacements for the birds grown on free-range pasture — birds marketed live in cities, before the machine age stole into the chicken coop.

The tale of the loss to family farming of the craft of chicken growing (it is a manufacturing process today) is worth sketching. It illustrates the vulnerability of the agricultural economy to the same processes that have removed first the skill and then the work itself from more obvious areas of industrial production, such as weaving, housebuilding, manufacture of furniture.

The stage was set during World War Two, when red-meat rationing introduced many consumers to the delights of chicken, and even spread the dominion of the victory-garden backyard flock. After the war the raised demand for chicken meat persisted. It was satisfied by many farmers whose flocks were a supplement to other agricultural activities, who marketed their animals side by side with professional — but still small-scale — chicken farmers. Birds were processed around the country in many thousands of small plants, and in butcher shops, and were locally sold.

The growth of the supermarket created a demand for birds

of standard quality, and at the same time, a nestful of technological developments hatched out with the potential to transform the chicken business. Antibiotics gave good control of chicken health problems, allowing investors, for the first time, to contemplate setting up large-flock, capital-intensive and low-labor "feeding factories." Mechanization of feeding, manure handling and watering the birds afforded a level of management standardization that allowed unemployed rural folk with no particular skill to undertake the rather custodial chore of raising chickens on contract. A business form, in which feed companies owned the birds, supplied the feed and veterinary services, and paid the new-style skill-free farmer a piecework fee, allowed the development of a business creature called the integrator — a corporate form that decentralized management decisions. What was once land-related, farm-related decision making shifted to hands whose main craft was keeping, astutely, at least one set of account books. These integrators took over and expanded a few of the many small chicken-processing plants in each area. The rest closed. It turned out that while there were some gains in thrift to be enjoyed by *raising* more birds under one roof than ever before, there were spectacular increases in the economy of *processing* more birds under one roof than ever before.

In a painful shake-out period that occurred during the late fifties and early sixties, the old farmers hung on as best they could, while the new "integrators," attracted by the opportunity to organize for high profit what had previously been a homely sort of activity, shouldered into the business as well. There were chickens galore, and they went so cheaply that the ancient promise of a bird in every pot was at last fulfilled. In the long run, this period of very cheap chicken served to swell further the ranks of poultry eaters. But it also broke most chicken producers, leaving only the biggest operators, capable of withstanding a run of many years in which the business, at the production level, did not turn a profit.

When the smoke cleared, the fox was gone but on the family farm the henyard was bare. Alarm clocks replaced roosters on many a back road. And one more opportunity, once taken as a matter of course — for supplementing farm income with a labor-intensive, quality small-scale enterprise — had become impossible.

Pigs are different from chickens. A different sort of management skill must be applied. With pigs, each animal wants notice and care. Variations between animals are greater. The number of person-hours required, with current technology, to produce a hundredweight of meat is far greater. Pig farmers usually raise their own feed, while broiler integrators came in on the capital of large grain companies. Large-scale hog butchers have evolved ways of keeping a flow of market hogs onto their assembly lines, while chicken processors usually suffered sporadic supply. Any hog farmer, especially Joe Weisshaar, will be pleased to testify that "hogs are stubborn," and the industrializers of hogdom are finding them so. But things are changing. Larger and larger hog farms are doing well. And larger and larger meat packers anticipate the larger farms.

The hog breeding business is changing in advance of the rest of hog enterprise, and it stands to reason. Not very many years ago, when hog farms were of one piece, the seams were invisible. Farmers who kept boars also bred them to sows, raised piglets, selected out a few of the fattening "gilts" to be the sows in his future, and went right on breeding pigs and selling pigs, litter after litter, lifetime after lifetime, for what must have appeared, to whoever bothered to think about it, to be a prospective eternity. When the seams of the business did become visible, it was because changing economic conditions, and an array of new technology, made different parts of the hog rearing trade differentially vulnerable to "improvement."

The part of the operation involving breeding, birthing, and nursing the young stock for the first couple of months of their lives was weeded out because it is so labor-intensive. What has happened simultaneously is that hog herd health problems, a marketplace structured to reward uniform lots of pigs, and the increasing size of on-farm batches of pigs, have made more and more farmers willing to pay top dollar for young stock possessed of commercially desirable traits — uniformity, a high meat-to-fat ratio, fast rate of gain, high tolerance of the stressed conditions of modern confinement housing, and certification that they come from a disease-free environment. Here, at the beginning of the business chain that leads to the marketplace, "corporate farming" has forged strong links with the hog industry.

Most hogs are still raised by relatively small farmers operating in relatively "pure competition" situations. But most breeding stock, most of the on-farm boars, even on small farms raising piglets, are not far removed from a decreasing number of major breeders. The reason is that good breeding costs so much development capital. It involves raising and testing large numbers of animals, culling all but the best, and going at it again for many generations. It involves labor-intensive processes all along the line, from artificial insemination to feeding test animals measured rations and recording their rates of gain. It involves record keeping and salesmanship. All this takes place, nowadays, in a world of big business and high finance removed from the everyday world of Corn Belt hog raisers.

The results of this sequestering of genetic material are quixotic. On the one hand it produces meat thriftily — average rate of gain per pound of grain is up. In a world where food is scarce, moral congratulations are due breeders of an animal that gains weight on less, just as there is something to be said for cars that go as far on less gas. However, there's little reason to think that the grain thus conserved ends up in the possession of the formerly starving, just as the gasoline not consumed in small cars is not spent instead on socially redeeming energy use.

The newest wrinkle in the hog business is that the largest companies are buying up the already large producers of genetic stock. It is the point in the production chain at which each investment dollar buys a great deal of control; breeding stock is the constriction in the flow of pork from the Good Lord to the marketplace. When few enough companies control it, they may find it handy to stop competing with each other, to set the price of thrifty genetic stock at above free-market value.

For all his farming skill, decisiveness, organizational get-up-and-go, gentle generalship and lowliness, Joe hasn't yet been able to martial the capital to go ahead with the farrowing house plan. He frets, and he keeps on trying — writing letters to far-away banks, meeting with the farmers who want to go in with him. Once, a farmer as restless and ambitious as Joe would have ended up more or less inevitably with a farm that revealed his strength. That's no longer true. Now that everything is capital-intensive, financial barriers block the road to self-improvement. Before a farmer's determination to succeed can be manifested in new buildings and fat livestock, the tide of the agricultural economy and of the world political scene, and the mood of the local banker must all conspire that it be so.

Increasing technological and financial integration with the world beyond the farm gates has cost farmers a flexibility they once enjoyed. Unlike his father, Joe is constrained to stay on the road down which he started his agricultural life. He can't switch away from livestock feeding — he has too much equipment, too many buildings in it. He hasn't control of sufficient acreage to make it grain farming. Joe can't even wait out a year very easily without raising hogs, even when he feels the market will be unrewarding. There are, eternally, payments to meet on land and "the facilities." Cash flow, and therefore continuous production, are the important things. No more hibernating. Joe fills his hog house, cycle up or cycle down. Occasionally, even in Iowa, one does see a farmer who misunderstands the temper of the times, or who just stops caring. But it is unusual. Joe had one demoralized neighbor, adjacent to one of his pieces of rented land. Joe had to fix all the fence. On the other side, rusty machinery, heaps of uncoiling wire, falling storage sheds, and a few scrawny beef cows were scattered across a hilly and unclipped pasture.

In Iowa good neighbors make good fences. Robert Frost each year would meet his neighbor at the fence and together they would "walk the line / And set the wall between us once again." The Creston ritual is different. "The righthand half of the fence as you face it is yours to keep up. The lefthand half as you face it is the neighbor's," Joe explains. Even neighbors who let their own interior fences go, who leave their tractors exposed to winter storms and the judgment of passersby still keep shared fenceline taut. Joe's demoralized neighbor moved away. Now the new owner's tenant there does his rightful share.

The ground is wet. The ground is eternally wet, it seems to Joe. It drizzles again. Sometimes his frustration makes him cross, although he has said, "Farmers learn there is no point worrying about the weather." We are on our way to fix fence. Joe says little as we ride.

In the pickup are a little crank-up machine called a fence stretcher, a spool of wire, posts, a bag of staples, a hammer, a bucket of electric fence insulators. Joe has made a list. The truck leaves the road, heads over rolling pastureland, clipped back after a summer of grazing. It sounds like we're driving on a wet road.

Just a few hours across the Missouri River into Nebraska, the average yearly rainfall tapers off from about twenty-four inches to about twenty-two. It happens along the Elkhorn Valley. Corn growing ceases. Milo and grain sorghum take its place. Farmers stop betting on a thirsty crop and switch their money to a thriftier, but lesser yielding, grain. It's one of those real world isobars, a line on whose opposite sides the economic truth told by weather, prices, and good sense totes up differently. It's not a clean line. There are conservatives who stop nearer to Iowa, and gamblers who keep trying deeper into Nebraska, but it's there. Joe's farm is far back from the line, but still near enough to a low-rainfall area so that summer drought is no surprise. The deluge now continuing is one.

The rain has put Joe's pastures into good shape. The pastures are cordoned off into ten- or fifteen-acre lots, so Joe can shift the cattle around and better manage the quality of what they eat. Each lot has a water source — a stream or a windmill pump and holding tank. The streams are all uncharacteristically full now. The older fences are still tight enough to keep a hog enclosed, although all Joe's hogs live indoors now. Hog fences are one of the greatest expenses in pasture-system hog raising, and the quadrupling of the price of woven wire fence a few years back precipitated a final departure by many diehard small farmers, and a switch to confinement by most who remained.

Joe's mother has phoned with word of a breach in the fence, and we are headed for one of Joe's smaller rented pieces. He grazed a herd of pregnant cows there most of the summer, together with their calves. Now they are elsewhere, and the grass has just started to regenerate. As we arrive, we see there are cows in there again. The place across the fence is owned by a veterinarian, who bought it as an investment. He rents it to a farmer — this one from the next county. There's seldom anyone around to supervise. It's clear where the cows are coming through, because one is clambering over the downed wire even as we drive up. The cow is half wild, and on forbidden ground is even more alert than usual. She sees us, hesitates, then backs off ungracefully.

We start by enlarging the hole in the fence, clearing away rusting strands of wire. We drive the half-dozen yearlings that have ventured away from the neighbor's herd through the hole. Then Joe patches the opening with three lengths of new barbed wire. "You'd be surprised how well the old rules still work here," Joe says. "This is the only section I've had any trouble with at all."

Returning from fence detail, we stop up at Grandma's. Although Joe has been married for twenty years, Grandma's house is still the one he calls "down home." His own dwelling, four miles back across town, is "up to the house." Grandma lives alone now, and spends days housekeeping. It's the same home in which she raised Joe and his brothers and sister, the same home in which she nursed her husband, Hans, five years ago, during his final illness.

She spots us driving up. As we stand outside in the heat, scuffing mud from our boots, she waves. She looks thin, frail. Her skin shines. She is in her late seventies, Joe tells me, and she has had pneumonia earlier in the fall. She's just begun to feel strong again. We stamp into the mud room. The door to the kitchen opens. Grandma smiles. Pie smells waft through the doorway. Her pet Doberman — the whole family keeps Dobermans — dashes out, eyes me and greets Joe. There is a pie on the table, and a steaming pitcher. I smell coffee.

Joe says, "You've been busy. Baking pies."

Grandma says, by way of explanation, "I've just got to be doin' something. If I had to sit all the time I'd be like a coon on a chain! I haven't been up to keeping after the garden," she points out the window, "but I get the housework done, and there's always some grandchild or another here after school" — she points into the living room, where toys and kids' books are piled — "so I manage to keep busy. I've been something of a pack rat all my life. There's plenty of stuff I've got to sort through, put in order." She reaches for a scrapbook, pulls out a full-page feature from the Creston paper, interviewing not her but her brother Pete. Pete attests, in his eighty-third year, that he is still bartender at the Elks Lodge in the center of Creston, and that he still picks up his fiddle once in a while and plays "Chorus Jig" and "Goofus." Grandma takes back the article, hands me a plate of pie, and says, "He left out the hardship parts. Sit down.

"You take this wet fall," she continues, "I've seen worse. It was so muddy, harvest time — thirty-five, I think — you had to bend down a cornstalk and step on it while you were cutting off the next ear of corn so as not to sink in; we used to cut the ears with a palm-hook then. Whole crew of us doing it, just so's we'd have something to feed the pigs. Now, it's just Joe and the combine."

"If I had to farm like that," Joe says, "I'd quit."

Joe is settling in to down two or three slices of pie. I recall Mary Jane's emphatic comment on the subject. "He cares *so* much for his mother, so very much, it's always impressed me."

"I like to talk now. Never used to have much time for it," Joe's mother says. "My mother-in-law couldn't talk for the last three years of her life. Had a stroke. My, it was frustrating to her! Once Hans said to her, 'Don't you wish you was young again?' Her answer was, 'No . . . and go through what I went through again? Not on your life.'

"I've been through a little bit myself. I remember the year — I was young — the house burned down. People in the town got together and helped us, and we all pulled an old schoolhouse onto the land. We lived in that. This house seems bare to me now, even though it isn't. We lost familiar things. Antiques.

"The antiques were things the family had used, earlier on. When I was a kid we had a self-sufficient farm. Storage pits for root crops, apples. Hay, corn, and wheat is what we grew. My brother Pete made three dollars for fiddling at a dance then — owned a grocery and tavern later. When I was a girl still, we were so poor I used to go hire out to do farm work. And you know what? I worked for a woman. A woman who farmed then. Mechanics are a good thing to do for women. These used to bale hay. I'd help in the fields, and I'd watch the kids too. I'd get two dollars a week, and glad of it. I'd clean for that too. Now my daughter Jane's maid comes in nine and leaves one in the afternoon, and she gets twenty dollars. It took me a long month to earn that!

"I wish I had kept a diary so I could tell you all the things that happened. I have old checkbooks — and that helps me remember lots of things. This modern farming — it's too bad they had to go into it like this. They want to make it more corporation farming.

"Anybody like Joe — likes to work in the dirt — they can't buy anything now if something breaks without paying a terrible price for it.

"It's just out anymore — there's no social life anymore. No neighbors. The land changes hands. The people who bought Grandpa Weisshaar's place sold all but twenty-seven acres again. Land changes hands all the time. The big home we used to live in was just torn right down.

"It started in the thirties. *Everybody* lost farms. Bankers bought land. You couldn't borrow a nickel to save your soul. Hans' dad almost lost his farm for want of five hundred dollars. He was lucky, though. A banker he knew made an exception and lent it to him. A fellow named Mr. Hansen came in from Nebraska about then with fifteen hundred dollars and bought up three farms, just before land went up, including Hans' dad's place. After we sold it, we still farmed it.

"It was the Depression. Corn was fifteen cents a bushel, and pigs, two and a half cents a pound. We sold pigs for five dollars apiece. We paid the note off at the bank, though. Finally, another place we farmed got sold off — a man named Smith first bought it. We rented it and kept farming it right up until last year. Now someone else rents it. It's the doctors have bought up a lot of farmland. They're the only ones just about can afford it. This Smith place sold for a high price, after Joe'd already plowed it for the year. He's short feed now, and his brother John is out of the farm too. On TV there was a man who farmed — needed a certain section to keep his farm convenient, and he bid three thousand dollars an acre for it. Didn't get it. It went for three thousand five hundred. With this going on I sometimes wonder how's Joe, and how's his kids, going to make a living?

"When land was cheaper, there were a couple of times Hans could have bought some, and of course he did buy the home place — the one Joe owns now, eventually. There were others he didn't buy. Toward the end of the Depression, Hans had four hundred dollars put away. He could have bought land he farmed with that. He thought about it, and he went out and bought his first tractor instead. They'd offered the land. But just lately, he'd had such trouble with horses.

"People think it was fun to farm with horses, but it was aggravating. It was thirty-four, I think, and one horse got shot by a hunter. One got sick — a disease was going from farm to farm that year. We kept six horses, and two went that one year, so the tractor looked better than the land. Besides, it didn't seem like you could pay for the land with the prices you got for what you grew, and you couldn't borrow anything still. It made sense at the time; now I wish we'd bought what we could.

"Still, the tractor made it easier on Hans. He kept adding little pieces of equipment in order to do more with it, as soon as he could afford each piece. Even so, those were awful years. In thirty-four, chinch bugs took all the crop, and there was a drought. In thirty-five it was wet and there was no crop. In thirty-six, drought again. Those were the years of drudgery. In thirty-six I got pneumonia. The doctor came every day. Charged a dollar a trip and fifty cents for medicine, and it was hard to pay.

"I've always had health trouble. When I was eighteen, I used to love to go to the dances. It was always waltz time then. My brother was playing fiddle at one. He caught smallpox from a guy came up to the dance, and we all caught it from him. I think that weakened me, but I've always worked hard.

"We had cows — milked twenty of them. Used to cool the

milk in a hole in the ground we called a cave. Worked fine, but in the fifties sometime they passed a law you had to cool the milk in a fancy stainless-steel 'bulk tank' and it cost so much we got right out of dairying. We grain-farmed a little right along, and Hans always kept pigs. I did the housework and I milked the cows."

Joe finishes his third slice of pie and wanders out to check on pigs. Grandma Weisshaar pours me a fourth cup of coffee. "Where were we?" she asks.

"You were saying you did farm work," I reply.

"Never did farm work."

"You said you drove the tractor sometimes. Also that you used to put up hay with the women. Also that you milked cows."

"I did milk cows — all my life, as far as that goes."

"Isn't that farm work?"

"It was Hans' farm. Mostly I kept the house. There'd be a crew come with the steam thrasher that came by every harvesttime. You'd cook for fifteen-eighteen hungry men that had all worked. Meat, potatoes, gravy, all kinds of vegetables, and always desserts — lots of desserts. Four or five women cooking in one place. You just bet you'd be careful how you treated your neighbor. You needed him the next year. Today you hardly know who your neighbor is, and as soon as you find out, someone else moves in instead.

"Everything's money anymore. I think it's greed. It's the government, too. Dad was a Democrat. I am, too. We'd split our votes, though. When I first voted it was bad times. Harding was in. The Ku Klux Klan was around. You'd have one of them on your neck. They thought the Pope was going to come here, take over the country. I think sometimes that things would be better if he had.

"What's come of it? People aren't satisfied anymore with what they've got. If they aren't they just get a divorce and try something else. I sure pity the next generation. They're not used to work, hardship. Both work at a job and they get what they want too much. Being away from home when you have small kids — that's no good.

"Farm families aren't like that. Joe's a very good son. I'm proud of him. He's always thinking of me — and of his family. Joe is just like Hans — always there to help. When Joe was younger, he played football at school instead of working. He was a star. Then later, the G.I. Bill put him through school when it meant something to him. Joe was a Boy Scout. A 4-H Club member, raised calves. Always on a team. Now he's using what he learned. He knows how to keep up. He's so busy all the time. It's like my brother said about Joe, 'He's a damn good farmer, but a damn poor housekeeper.'"

Joe stamps into the house again. As if to prove his mother's point he stands in the kitchen while patches of mud fall from his pants to the floor. Grandma Weisshaar wraps up the rest of the pie, wraps another one besides, and hands them to us as we depart.

Joe begins his presupper routine of chores. He checks what's been done by others. No one came home from school to walk beans. He sees to it that what is still to be done won't suffer from still further delay. He looks in on all the pens and pastures, barnfuls and feedyards of animals, swine and kine alike, assuring himself that all can pass twelve hours unattended without losing him any money.

The cattle in pasture outside at Grandma Weisshaar's look back at us with appropriate late-afternoon placidity. We look at the pigs fattening there. They are too busy feeding and back-biting to bother looking back. Joe locks up the gas tank in the dooryard.

We load the two pies into the pickup and head back across town, "up to the house."

More pigs to check. The hog house is underpopulated since the trip to market and the reshuffling of lots, but more residents will come in a few days, after the emptied pen has dried out and been disinfected. Joe peers into the nearest of the hogs' feed bins to see that Dave, the farmhand, has done his part of evening chores before leaving.

"He always does, unless I've sent him off somewhere and forgotten about it," Joe says. Al sits on the cement floor of Joe's shop, taking apart a broken mowing machine. Joe affectionately rests his hand on Al's shoulder.

Mary Jane is in the house cooking for the third time today. I am the bearer of one and a half of Grandma Weisshaar's special pies. Mary Jane says to set them down over there, on the counter, next to the microwave. The microwave ticks and whirs. It is the diamond in a kitchen bejeweled with appliances. Moving clockwise from the microwave, there's a four-carburetor toaster, a thing that transmutes garbage into hard blocks, an innocent-appearing bit of counter called a "cooking surface," an electric beater, electric bean pot, electric egg poacher, electric can opener, electric dishwasher, electric blender. The microwave is like a TV. It lights up when turned on. It does things that go against nature — food cooks in it from the inside out. Mary Jane says Joe finally gave her this, after she'd wanted one for years.

"That is, he tells me what my gifts are. I'm a bargain hunter. He knows I always shop around. After it arrived, I got to like it. I read the stuff about the waves being harmful, but this brand is O.K. I had a dream about it, though. It was a very strong dream. I remembered it for days.

"In the dream, I went into the kitchen and the microwave oven was on, but the oven door was half open, which of course can't happen really because there's a safety lock on the door that keeps it from opening when the juice is on. There was a lion inside. I didn't know what to do — which was the greater danger. I didn't know whether to go over and shut the door because of the radiation, or run from it and perhaps be followed by the lion, because he wasn't locked in."

A bell rings inside the microwave and there is a loud click, the same sound that comes from a pinball machine issuing a free game. The light goes off, the door swings half open. The odor of roast pig, not lion, fills the room. "It's one of ours," says Mary Jane. "I hope that the boys come in soon, while it's hot."

We both glance out the kitchen window, checking on the progress of evening chores. Joe and Al have come out of the shop. They are standing together chatting, leaning against one of the head-high tires on the combine, which is still parked by the driveway. Joe has his arm around Al's shoulder. Al leans against his father. After a moment, Joe glances toward the house, and Mary Jane beckons violently, using her whole forearm and open palm. Joe nods his assent and smiles, but neither father nor son moves.

"That's what I like about farming," says Mary Jane. "The family is in one place. Your kids aren't running around and you not knowing where they are. Joe is a real natural with the kids. He and Al get along as well as any two people can. And Julie — maybe 'cause she was the oldest — Joe's always doted on her. I know it's normal, but it took me a long time to realize I was jealous of her in some way. He just liked her so much and I've never got along with her real good. But now we talk things out — it's a very close family. Jeannie, the little one — she's my baby in a way. She's always with me.

"Joe talks about everything with me. When we were first married, he didn't so much, but I think he just took after his dad. Hans was a wonderful man, but he didn't say all that much, and it took Joe a while to learn. I was always half afraid of Joe's dad. He growled when he talked. I remember, the first meal I got him, he wanted horseradish and he wanted mustard. The mustard hit the floor, and it sprayed at him, the ceiling, next room even. He didn't say anything. Just cleaned it off himself and put some on his hamburger. I felt pretty bad about that, but later on we laughed about it. I always thought Hans was tough and steely, but his daughter — she says one time, it was when she was getting married, they got into the car to take off on their honeymoon, and then she says, 'I got to get out of the car and say goodbye to Daddy,' and when she did that she kissed him, and she saw that he had tears in his eyes.

"Joe's gentle like that, but it comes out more — I brought it out in him more over the years, made myself more a part of things than maybe his mother was down to home. We switched our beef from Angus to Charolais a few years back, and the first lot of them had bigger calves than the mothers were meant for, and that made difficulties. Lots of nights we'd both be out there. One night he was pulling out a calf — had the mechanical puller on it and everything, and I started laughing, said to him, 'Sure glad you don't have to deliver my children.' But we were out there together. We got quite a laugh out of that one. He's humble in political meetings with other farmers, even when they are slow to understand something that is clear to him already. He's very good at political meetings. Another thing — he used to hunt ducks, but he gave it up. 'I feel sorry for the birds,' is what he said. And he's always with his family. He cares so much for his mother."

With this observation, Mary Jane bravely opens the door of the microwave the rest of the way, Joe opens the door of the house, Julie and Jeannie arrive from their stations in front of the TV, where homework was being done intermittently. Hands are washed, seats assumed, grace recited, meal consumed, including all of the one-and-a-half pies. Al returns to the shop in the barn, Julie and Jeannie return to TV and homework, overseen now by Joe, who always catches a brief evening nap in his recliner.

On the TV is a TV lamp, cast of ceramic in the image of a comic owl. Julie sits under a macramé wall hanging, also of an owl, and Jeannie, who is restless and bored, moves back and forth between the left end and the right end of the long couch across from the TV. On end tables to the left and right of the couch perch small white statues of owls.

In the kitchen Mary Jane loads the dishwasher. We talk about her machines. "They do save me time, and I use the

time for other things. And that's what the owls are all about.
They're my emblem, my political emblem. First I helped my
neighbor Roger Blobaum run for the legislature — he lost, but
not by much."

Roger Blobaum is not only a neighbor of the Weisshaars but
is also the elder statesman of what has come in recent years to
be known as the "ag reform movement," a loose network of
small research groups, lobbies, and organizations with regional
action programs, working to keep family farmers farming. It
was Roger who had first led me to the Weisshaars, saying about
Mary Jane, "She's a very skillful and talented organizer. Very
issue-oriented, and very hardworking and original. A few years
age there was a local race for national delegate to the Democratic
Convention. She ran against the Democratic District Com-
mitteeman. She had it organized, worked harder than you can
imagine, and beat him, an upset victory."

"What the owl is all about," Mary Jane says as she puts the
last plate into the dishwasher, "is that it was part of my
campaign slogan there. I had stickers all over. They showed an
owl head, and they said, 'Wise Who??? Weisshaar.'

"I've also been active in the National Farmer's Organization
for a few years now, along with Joe. They focus on couples, in
a way, and I've done everything. Spoken, taken cheese around
when it was part of the program to help out a cheese co-op
factory. Once Roger got us on TV. A crew came here doing a
show about farming in America. They were too much. The
crew was just dyin' to see — now let me see, how did they
put it — 'the mother pig deliver a baby.' That was it. They
saw it, too.

"Carter's Inaugural Ball — it was quite a night. Got pinched
there by a man I never saw before! Here, I'll show you something broke my heart!" Mary Jane rummages in a cabinet, sits
me down at the table, and opens a scrapbook to a portrait of
herself, with Jimmy Carter standing next to her beaming. She
had worked hard, became *someone*. This is the authentication.

It's blurry. The cast of characters is easy enough to identify, but the picture is fuzzy. Mary Jane seems forlorn, years after the event.

"We didn't bother to have it blown up," she says, "because it came out so badly."

She was back in the news again early in Carter's presidency. Joe had been invited to Des Moines to a breakfast with Jimmy and some farmers. Mary Jane wasn't included. She claims a matter-of-fact statement of hers was taken for a militant stand, and that she didn't really mean what she said "ironically." But when a reporter, who had heard via the ag grapevine that she was perturbed about not being invited, asked what she'd do, she said, "If they won't let me in, I'll just wait in the car." It was in the Des Moines paper. They let her in.

"Women's lib," as it is still known in rural Iowa, has touched Iowa City, Des Moines, and even Ames, but appears to have passed over the countryside. In Iowa, most women are still born to wed, bake, and feed the baby. An army of "farmers' wives," concerned enough with the political base of their economic well-being to have formed a rather fierce lobbying group, nevertheless puts itself down by calling itself "The Porkettes." It sounds more like a name for a kickline in a fairtime farce than a name suitable to a group that takes its politics as seriously as Iowa women do take pork politics. Who can conceive of lobbying welfare mothers calling themselves "Welfettes"? Or the irate wives of West Virginia coal miners calling themselves the "Minettes"?

What is indicated by the diminutive term "Porkette," by Mary Jone's self-effacing irony, "If they won't let me in, I'll just wait in the car," and by Joe's mother's modest belief that she was "not a farmer" even though she'd spent her life doing fieldwork and milking cows, is an attitude that has endured long past the era when it was a social requirement.

After the Civil War, farms in America ceased to organize as homesteads vending surplus goods, and began to grow food for the marketplace. The work of men could then be measured in cold cash. Women's labors on behalf of a "household economy" continued to be valued only outside the marketplace. Men made money; women spent it. (This line of thinking from Ann D. Gordon, Mari Jo Buhle, and Nancy E. Schrom, *Women in American Society: An Historical Contribution.*) At the same time, the broadening of social services such as education and health care diminished the dominion of "household economy" still further. Conveniences, from running water to Mary Jane's electric kitchen, further decreased the amount of women's work. Lessening family size narrowed it still further. The net result was an increasing disproportion in the relative value to each family's survival of the sorts of work done by men and by women. The transformation illustrates the complex consequences of some advancing technology. A multitude of machines billed as "labor savers" and increasers of efficiency did indeed end, for both sexes, the age of drudgery. But the machines, by disproportionately increasing productivity of marketable goods, devalued household labor without ending the demand for it.

Mary Jane is not a member of the Porkettes. "They do good work, but I am involved in so many other things. I'm on the board of a child care center here in Creston. Back when the NFO was selling cheese to help out a co-op cheese plant in Wisconsin, I was out there with a truck as much as I was home. I've started to sell office machinery for a local company. And I'm involved with the 'right-to-life' movement here. I see by the look on your face you are probably on the other side of that one. When would *you* say the soul enters the body — at birth?"

The farm couldn't run well without Mary Jane, whose household work, needless to say, keeps everyone fueled, clothed and clean, whose peak-load farm activities, from trucking to combine piloting, provide skilled labor, and whose affection and community leadership set the moral and political tone of the household. It is also obvious from Joe's demeanor and words that he would feel no point in farming without a family to "farm for."

Mary Jane's outspokenness, verve, and independence mark her as far less traditional than most women on farms in Iowa. Mary Jane is trying roles and making demands that are setting her apart from more conservative farm women. The coming of farm technology that left Joe's mother without a role to call her own has encouraged Mary Jane to try her wings in the outside world.

These same technological advances have forced Joe to trade independence for participation in a market economy so complexly integrated that he is increasingly forced to specialize, to become an element in a countrywide, statewide, even nationwide production line that by its mere existence determines how his next dollar must be spent and what chores he will do in the next working day.

If laborsaving technology and the world of big business have removed from Mary Jane the possibility of filling an urgent on-farm position, they threaten to do the same with Joe. More and more of his farming time is taken in managing costly inputs. Unlike farmers, managers are made, not born. They are interchangeable. They substitute regularity for wit, usual procedure for adventurousness, dutifulness for competitiveness, and obedience to policy for independence. They replace skill with system and accept corporate goals in place of goals that express personal spirit. In short, what farmers do, and what managers can't do by definition, is exercise craft.

Loss of craft in farming is serious, not just to farmers, but to the nation. It is the step before loss of pride, loss of personal ethics in trade, loss of stewardship of the land, loss of concern for quality of product. The loss reverberates all the way down the food supply chain. It can be felt at McDonald's, and in the aisles of supermarkets. It is part of a grander loss yet, the dying of a system of people making money doing things well. Supplanting the old system is a new one with slots for people to do what is prescribed. If farm women face a world that is sexist, farm people in general also face a world that is increasingly anti-individualist. If women count for little, so do we all, and the fights that Joe and Mary Jane in particular face are struggles against the same corporate and technological forces that trouble us all.

There's a strong sunset tonight, and despite my childhood notions to the contrary based on sheer Yankee ignorance, it is obvious this evening that Iowa is one of the most beautiful states in America. Some of Iowa is flat, ironed into squared vistas that could appeal only to sensibilities trained since childhood to adore them. The sections of Iowa I find so special undulate. Iowa undulates in its northeast, northwest, and southwestern parts. It pitches up and rolls down about as steeply as farmland can pitch and roll while remaining navigable by tractor. And that is the secret of its beauty. Its loess hills are made not of rock, like Yankee hills, but of good soil.

Iowa roadbuilders show great regard for the farmers who work the hills, but no regard at all for the hills. Fields are squared — farmers almost never have to plow point rows in Iowa, nor weave their combines in awkward figure eights to fetch up to that last little row of corn on the brow of some roadside hill. The fields are foursquare, and the roads never cross them. From the air the view seems flattened. It's as they say — Iowa farmground does look like a patchwork quilt. From the ground, from the crest of one of Joe's hills, the vistas still resemble a patchwork quilt. But from the ground, the hills this evening nestle, climb and fade downward into shadow. It's like a child's bedding-level view of a patchwork quilt, rising and falling, draped from horizon to horizon over a bed occupied by sleeping parents.

This is my fondest memory of Iowa: Joe drives a red tractor across a green hill, painting the ground with the tankload of brown pig manure he draws behind him. He is far across a ravine and up a steep slope; his tractor makes only a blurry whispering sound, like the purr of a pleased cat. The children Weisshaar, Jeannie, Al, and even Julie, who almost feels too old for such things, cry out in boisterous shrieks of delight. They are behind the house, on a platform high up in a tree. They fly through the air, one after another, dangling by their arms from a trapeze that rolls on a pulley down a long wire cable to a pile of mattresses fifty yards away. They shout me up the tree, thrust the trapeze into my hands, and wait patiently while I take measure of my fading youth.

The ride is wonderful — longer than I'd expected. It is filled with separate events — the resolute takeoff, the quivering of the cable felt through the hands, the view of the green and golden acres of the eastern slopes, lit by the sunset, and of the brown acres to the west, in the shadow of twilight already. Then the landing, the nestling down in mattresses which turn out to be wet, surrounded by the memory of the patchwork quilt I tumbled onto when I was four. I look up, shaking myself back into this world. Mary Jane is standing at the base of the tree, laughing. The guest has survived, and pledged himself a mission to return the favor.

In one way, the beauty of the rolling land of the Corn Belt is deceptive. Its extraordinary fertility is endangered. Soil scientists now tell us that for every bushel of corn that is taken from the hillside fields of Iowa, a bushel and a half of topsoil washes away and another half-bushel blows away. Two feet of topsoil lay under the native prairie sod of Iowa. Only half of it remains. What has been lost is gone because it paid best to farm poorly. We fancy that farming is a "renewable resource" — that, when we are out of oil and uranium and iron ore, still there will be land to farm. It's not true. The current thrust of mechanization threatens not only forests and fish and oil reserves, but the very soil as well.

It has to do with bigness — big machinery wants clear ground to do its work. In the past half-a-dozen years the once spectacular progress made by the soil conservation service, following the dustbowl days, in introducing strip cropping and contour farming, has to a large extent been eradicated by the press of short-term economic need. Big combines turn awkwardly. Six-, eight- and ten-bottom plows need room to swing around and get started again. If every field a farmer plows must have wide end aisles for maneuvers and roads along its sides, it obviously pays to remove obstacles and join fields together. Tree lines and fencerows — which are good windbreaks and good harbors for predators of plant pests — are each year a rarer sight in Iowa fields as farmers choose larger tractors to do their work. The farmers do not join fields together blindly. But their choices are forced. They know, says Joe, that what they are doing fosters erosion. They say it can't be helped.

These are the same farmers who all across the Corn Belt signed up eagerly in the forties and the fifties for Extension Service help in laying out contoured rows that wind crops around hillsides, interspersed, in the best of cases, with strips of unplowed grassland. These same farmers feel grateful that the conservation movement afforded them the opportunity to avoid another Dust Bowl. But today they plow straight across the same fields. The patchwork runs truer — less crazy-quilt, more businesslike order. In the long run, though, it may be bad business. When short-term demand makes the squandering of resources profitable, resources are squandered. It's a mechanical glory of the system. Farmers farm as their situations dictate. These days more than ever before, they are likely to plant corn on their flat land, and then when they run out of flat land, to go ahead and plant corn on hill land that shouldn't have row crops on it at all.

The result is soil loss. Iowa conservation officials estimate that hill land now loses about thirteen tons — two hundred and sixty bushels of soil per acre, per year. Farmers counter the fertility losses with increasing doses of fertilizer. Because of the trend in recent years to plant "continuous corn" — to follow corn crop with corn crop while making up for decreasing soil fertility with fertilizer — fertility loss from erosion has been masked by fertility gain from incorporated fertilizer. Even at current prices, it pays to load good land with all the fertilizer crops can bear. Herbicide use — vital to a continuous corn program — accelerates erosion from fields barer than they've ever been before.

Hilly fields wash worst. Yet farmers have been plowing and growing crops where they would never have dreamed of doing so just a few years back. The Soil Conservation Service (according to *Successful Farming* magazine) says that in 1976 alone, United States farmers converted nine million acres, most of it previously in grass, to tillage land. It is not hard to understand why. Cropland pays a better return — as long as it stays put.

The rare farmers in the Midwest who own all of the land that they farm are in the best position to keep on being kindly stewards of the land. Like Joe Weisshaar, most farmers in Iowa nowadays rent some of the land they farm. And that makes them vulnerable to the current economic crisis. In the past five years, the value of the land that Joe Weisshaar owns has doubled. The three hundred fifty additional acres that Joe rents or "farms on shares" have similarly appreciated in value. While Joe has had the privilege of deciding how to deploy the home farm, he hasn't the same control over his rented ground. Rental prices rise as land prices rise. Joe's landlords have constantly to choose between selling out in order to capitalize more rewarding investments, and keeping their land because they feel it makes acceptable returns. "I had one landlord once used to come to me every spring, in the *spring*. He didn't know what the weather would be like all summer any better than I did. And he'd ask me, 'How much money am I going to make off the land this year?'" Joe says.

The idea — and it is one lodged deeply in our baggage of cultural tradition — that land is an investment, a private commodity to be traded for any purpose that strikes the owner's fancy, is at the root of the problem. It seems not to be something that is about to change this side of the Revolution. (In fact, it's the basic issue in most revolutions.) Competing demand for investment capital makes it inevitable that landowners require their investments to yield typical returns. This used not to be the case when landowners farmed their own land, and when land was cheap.

Nowadays, landowners across Iowa, the knowing and ignorant alike, feel forced to tell the farmers who work their land, "Joe, you know that fertile hillside that sits in pasture? Move off the cow herd and rotate corn and beans on it. Plow it and plant it."

The economics of corn looked like this in 1977. The average Iowa acre of corn returned 88 bushels, and the bushels sold for $2.30 each. Each acre, then, grossed $202.40. Farmers estimated tillage, planting, cultivating, and harvesting costs — including labor and machinery costs — on an acre of corn to be about $125. If the farmer rented the land on half-shares, which is common, then the landowner took $101, to pay the mortgage, to invest in nonagricultural pursuits, to do as he saw fit, to compensate for keeping money in land and not investing it in a shoestore instead. The farmer broke even if he was a thrifty operator. He didn't get rich selling corn. If he fed the corn to the pigs, he ended up making $50 more on the corn (if you assign all the profits to the corn, none to the pig or pig-rearing labor, which you wouldn't do). Had the same land stayed as hill pasture and grazed cattle, both costs and returns would have been lower. If the landlord is ambitious and restless, he will shop for a farmer to rent his hill land who is willing to plow it and plant corn on it. The landlord then takes home a rather handsome rent. On the other hand, the farmer does about as well on beef or corn on rented land.

Between February 1976 and February 1977, the price of a farmable acre of Iowa dirt rose 15 percent — the value of a suburban house lot in Connecticut only rose 8 percent in the same period. It keeps farmers like Joe, farmers who have not inherited modern-sized farms but have had to capitalize them themselves, on the edge of a rather posh subsistence. Whenever Joe has a good year he feels compelled to turn the profits back to acquiring more land. But his competitive position is not strong.

"I tried to buy the place back of Mom's," Joe explains. "There it was — pretty good land that Dad and I had farmed. The fellow said he wanted twelve hundred dollars an acre for it. It sold in a little while to an accountant in Des Moines, who rents it out now."

Joe is in the same position on other pieces he farms. He even rents the farm on which he lives. "It's not for sale. I do get better terms from my uncle than I'd get from any little old lady with a farm management service company handling it. Still, he takes half the crop."

Every time Joe works on a rented piece of land, he has cause to regret decisions that seemed the right ones to make back when he was fresh out of the navy, newly married and just starting to farm. "I was always being too cautious not buying land back then. That was my biggest single mistake."

Not only accountants but also some farmers are buying up the expensive farmland, and making good business sense out of it, too. These are the largest and best established farmers — the "strong hands" to whom banks extend cheap credit, who own sheds full of the largest equipment available. For these folks, additional land to till does not always require additional machinery investment. The biggest farmers have land-hungry equipment. They can't afford to stay small, any more than the stable farm with equipment contoured to the operation can afford to get large. It is a case of two management philosophies colliding, and it turns out these days to be a collision between a bus and a compact. What's more, the large farmers are also in the best position to pay the top-dollar cash rent demanded by outside investors in farmland.

If Joe is forced out, he will not go out poor. Farmers do make little, relative to the size of their investments, and they are fond of getting on TV and waving their income tax forms about to prove their poverty. But if one adds an annuity to farmers' yearly salaries equal to their annual increase in net worth due to rising land prices, equipment values, business assets, et cetera, it is easier to believe that farmers are not economically irrational. On paper, Joe made nothing last year, on a net worth exceeding a third of a million dollars. Even so, his net worth groaned upward just because his land grew more valuable.

Yet his enterprise is far from solid. He is threatened by the high price of land on several counts. Not only is he forced to farm in a manner that fails to preserve irreplaceable soil (the loss here is a slow one, a threat to Joe's children), but he is forced to compete for every piece of land he farms with the growing array of nonfarming land investors.

More and more, the ownership of land is becoming separated from the farming of it. Farmers are becoming nonproprietary businessmen. Large farmers successfully outbid smaller farmers for land that does come up for sale. Accountants, veterinarians and smart bankers with an eye for future values are in the running too. And still another element, foreign investment capital, has recently entered the farmland marketplace. Foreign investors, from Europe, the Near East, and from Asia, attracted by the depreciated dollar and cautious about political instability at home, find U.S. farm acreage a safe haven for spare cash. Knowledgeable farm real estate people estimate that overall, a fifth of all the farmland now sold in the U.S. is sold to foreign buyers, and that in certain prime areas of California and the Midwest, the figure is nearer 50 percent of sales. In 1977, according to a large common market investment company, eight hundred millions of European dollars alone were spent on U.S. farmland. Many of the purchases are made in a manner that masks the identity of the buyers — purchases made through tax havens or complex corporate structures. Overall, the amount of farmland marketed represents a small percentage of all farmland, and foreign investment is not yet a big part of the national picture. To date twenty states have banned or limited foreign ownership — although legal experts say the laws are easy to get around. So far, foreign investment is a limited but significant factor in the total changing land-holding-scape, strengthening the trend toward nonproprietary farmland ownership.

A potentially more volatile development is the entrance of investment funds into large-scale landholding. This is hard to evaluate, because landholding statistics are hard to come by, and buyers frequently veil their consolidated identity by using many subgroups to hold land commonly managed.

In 1977, a group called Ag Land Trust Fund One, backed by the Continental Illinois Bank, put forward a plan to invest $50 million of customers' money in midwestern farmland. Only a dramatic public outcry, led by ag reform heroine Paula Schaedlich — a young Iowa researcher, who read of the plan and sent out a large personal mailing about it — caused the investment to be shelved for the time being. The tendency is there, however, and the capital formation equipment is in place for investors (both corporate and private) to buy in.

It's a practice that has kept generations of European "first families" in power through war and changes of regime. When the smoke clears after whatever national traumas or transformations occur, if the deeds are intact, land is a good base for continuing wealth. In some cases land may be so attractive a haven for capital that wealthy investors and conservative corporations are willing to accept lower returns on their purchases than local investors. This allows them to pay higher prices, to outbid regional purchasers in the current volatile land marketplace.

The trends are not yet well established. But it is already clear that change is happening, and that it is the sort of change that spells further troubles for family farmers such as Joe Weisshaar.

The trouble will be the detachment of landholding from land tilling, and if it is a growing problem on land that comes up for public sale and is acquired by these foreign, large, or institutional buyers, it is also a problem on the sleepy backroads farm that does nothing but change hands within a family. The trouble is a direct result of suddenly increased land prices, and it works like this.

Grandpa bought a farm for $200 an acre in 1911. Say it includes 500 acres, and was a judicious purchase — Grandpa bought good land. Today, the land is worth a million dollars on the open market. When Father acquired it from Grandpa, in 1939, it was worth only $350 an acre on the open market, and Father was encouraged to acquire it all himself. His two sisters, half-brother, and surviving mother were treated somewhat shoddily elsewhere in the will, but the principle of keeping the farm intact was observed — the farm stayed alive. Today the stakes are just too high for such a transfer. The sort of settlement made when the farm passed from Grandpa to eldest son is no longer frequent.

The disparity between the great worth of the farm and the inevitably small value of nonfarm assets is just too large — family justice can only be done by dividing up the formal ownership of the farm, and it is happening all over the Midwest. Father dies. Son takes over — but no longer as sole proprietor. He is now certain to be a minority shareholder on his own farm. Let us give him two sisters who live in cities — one in Detroit and one in Los Angeles. We will also endow him with a younger brother who is an executive with a fertilizer company. Our farmer, as a shareholder, owns a fourth of the farm. The other three siblings let him manage it. He farms on shares, half for himself, half for the shareholders — so he ends up with five eighths of his own crop. Let's say this situation goes on for twenty years and he up and dies of heart trouble.

He leaves one son, eager to farm — also a wife, two daughters, and another son. Attorneys find that the sister in L.A. has also died (the heart condition runs in the family), leaving three 'kids — city slickers, every one. His sister in Detroit is senile (same problem) and her estate is being managed by a trusted banker who is being sued by the sister's sole heir, a petulant daughter. Meanwhile, the fertilizer executive has had a feud with his brother and doesn't like the son who is taking over, either.

What's left is not a family farm. It would be simpler to farm the land were it owned by an Arab or by the most cynical investment-oriented banking consortium. What this swollen example makes clear is that land prices have risen to an inconvenient level, one that will destroy the traditional intergenerational transfer of family farms. Farmers will keep farming, in the sense that they will make the immediate operational decisions about what to do on the land next in order to make money growing things. But the trend toward nonproprietary farming changes the very meaning of the word *farming*.

The concept that clarifies the implications of shifting control of farmland comes from a paper written in 1974 by an agricultural economist named Marshall Harris, and published, surprisingly enough, by the government — by the Economic Research Service of the Department of Agriculture. Harris says simply that what is changing, between the good old days and nowadays, is the nature of "entrepreneurial control in farming." He says what is afoot is a development of ". . . changes leading toward loss of traditional decision-making power among farmowners and an exercise of power of decision making among parties not holding conventional ownership. The trend is toward 'property without the power' and 'power without the property.' "

Harris argues that farmers used to have virtually complete entrepreneurship but have lost much of it. They had it when they had fee-simple ownership, tilled no rented land; owned and bore risks on all capital used; used family labor; had complete managerial control — how much to produce, what, and where to buy and sell; and when a crop sold in a free marketplace.

Nowadays, farmers live in a world where other business structures exist more powerful than themselves. Farmers' land is likely to be rented, at least in part — frequently in large part — from others employing a manager or management company to oversee decisions in the interest of landlords. Farmers buy expensive land and equipment, and finance it with loans that assure lenders a hand in farming decisions. In order to assure themselves access to controlled markets for some commodities (eggs, broilers, milk, feeder pigs, tomatoes, pickling cukes, canning vegetables, among others), some farmers enter into production contracts with processors or distributors that include sharing of control of farming operations. And, granting the great good done farmers by their cooperatives, joining a co-op also diminishes the independence of farmers' own choicemaking.

Modern times have forced farmers to allow shifts of some entrepreneurial control to landlords, output purchasers, suppliers, lenders, cooperatives, and even to the government. In return, the farmer can stay in business. But — and it may be news to those afflicted with that brand of romanticism described by a friend as "Currier and Ives syndrome" — the business that farmers are in today is, more than ever before, business.

A horrid little article graced the pages of the influential *Farm Journal* a few years back, written by the editor, Gene Logsdon, and describing "Hiram Drache, a keen analyst of farm trends who is also a farmer, feeder, professor and historian." Professor Drache, who operates on a large chunk of Corn Belt land in Minnesota, is quoted as saying, "Keep the people back in the cities and leave us free to farm. We'll send 'em food!"

Drache speaks with the kind of innocent forcefulness that is the gift of those who will gain from generally discomfiting change. "A virtual revolution in farm consolidation is taking place, the extent of which is evidently beyond the grasp of most farmers," the article quotes him as saying. "Larger commercial farmers are rapidly expanding by using high-priced labor, paid managers, heavy borrowing and huge equipment. Meanwhile small neighboring farmers, blissfully unaware, insist these very tools of growth spell the doom of large-scale farming."

Drache has studied trends in farm size (farms that make money are getting larger), interviewed farmers who manage large farms, projected technological developments that will permit increasingly productive use of human labor, and he envisions a future for farming that seems both unpleasant and quite possible. Drache likes it. I cringe to think about it.

> We're nowhere near the end of growth in farm size. If farmers feel that such large farms are technologically practical today, I think the farms of the year 2000 will be much larger than the average person today dares to conceive. I'm convinced that we *could* be down to 100,000 farms producing more food than now at a cheaper price. That's not outrageous. That's only 3,300 acres per farm.

Drache sees farmers farming more and more rented land. He envisions a nation in which there are many landlords but only twenty or thirty large farms to a county. And, amazingly, he persists in thinking of himself as a champion of family farming. "The farm of the future will be a family farm, but a bigger, more sophisticated family farm." He sees the largest farmers coming to be resented by their communities — he attributes the feeling to "jealousy." He speculates with seemingly calculated overoptimism that the surviving large farmers will not comprise a "farm bloc" anymore, because, he reasons, the political organizations of use to smaller farmers are not helpful to the big fellows.

This vision of the future seems possible, and one has to admire the audacity of Hiram Drache's insight in putting the trends together and saying so. What seems hard to believe is his notion about the resulting political power of farmers. The smaller the number of farmers, the greater their ability to coordinate actions resulting in imperfect competition and, therefore, in *higher* prices for consumers. It may be true to some extent (that is, if one ignores the influence of tax laws) that increases in capital outlay are only possible because they lower per-unit costs of production. But it doesn't follow at all that lower on-farm costs will be passed down to consumers, any more than lower per-call telephone costs resulting from economies of scale are passed on to consumers. It takes competition to insure this sort of pass-down.

If I were a big farmer, I'd set up a big farmers' club and whisper about controlling acreage planted, about integrating for greater safety, with suppliers and processors, in a form that reflected club muscle. In short, I would minimize free enterprise to maximize profit.

Says Drache, "The danger is . . . in consumerism and government intervention which can check the free enterprise it takes to solve problems. Political manipulation of the economy creates only a nightmare of bureaucracy." Our devils clearly blow in over different quarters.

I don't see enough people doing good work in Hi Drache's version of the drama. On the other hand, the "Ag-reform movement" has a deus ex machina in the form of an energy crunch that raises the costs of large-scale farming most, or a rising consciousness about health-related issues, or the impossibility of managing large farms efficiently, or the growing power of consumers, thumping down on stage in time to save our farms. The liberal thump of salvation is coming on stage rather softly. For all the unpleasantness of his vision, Drache may well be the soothsayer.

The attitude behind his pro–big business sentiments is an alluring one, one that allows relieved believers to feel that the unstoppable, inexorable powers that are moving the whole construction of society around are good, reasonable, just, as they should be, not to mention that they are assured and inevitable.

I wish I could think that way; it would be comforting to be affiliated with glory, to think the powerful possible, dreary, disabling and unwarm future just, appealing, meet, proper, more than we poor beggars deserve, a thrilling prospect that will keep us safe, a reassuring and enveloping development that *will happen*, if some dangerous, insidious, misinformed loud-mouth consumers shut up who should turn off, sit still, and stop ruining the future for the rest of us.

Near as I figure it, Hiram Drache's thinking, as reported in the *Farm Journal*, either reflects this sort of proud but obedient authoritarian attitude, or, less probably and less pleasantly, reflects a kind of disarming cynicism, a posturing as "farmer" by one who stands to gain when farmers are a thing of the past and a small race of Business Winners takes their place. In either case, what Drache argues is not perfectly likely. It seems an extreme version of something that is likely, however. It is true that farm rentals, the incidence of farmers as minority shareholders on the farms they farm, farming by managers, and farming by consensus in widely integrated multibusiness operations are all on the rise. But if we can preserve a relatively free food marketplace, the efficiency of actual farming operations will still count. As long as efficiency counts, efficient farmers will have a competitive edge. And efficient farmers are not usually large operators.

Farming is an unusual sort of enterprise because the density of management decisions remains so great, even at the production level. Farmers have to respond quickly and flexibly to changing crop conditions and marketing conditions and machinery repair conditions and labor conditions. Size inevitably breeds clumsiness — and the modern management structures that are designed to work flexibility back into stiffening systems themselves add layers of costly management and costly feedback operations that divert management time from management and that demoralize laborers who might once have been prone to singular acts of productive initiative. The *real* American way, the habit of commerce that historically has led to the sinewy growth of American capital, is simpler: people work well when they work for themselves. If we have passed the stage when commercial life can be organized that way — and we may well have — it seems appropriate to worry about the political power of the few who will provide us with our daily bread.

To date, farms are still bite-sized, still harbor craftsperson-owners, such as Joe Weisshaar, who can compete even in this age of bigness. The finesse needed to farm efficiently is an as yet unalterable agri-fact. Any diverting of farm systems from this straight and narrow will be diversions that cost consumers money and further take the craft out of farm culture. As long as the free market persists in delivering the goods farms grow, most sorts of farms will stay relatively small. They may be large for farms, but they will be small compared to other U.S. businesses that make things. That's what will happen if free market conditions for farm commodities prevail. Unfortunately, they may well not.

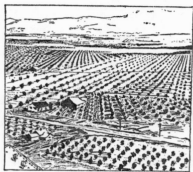

The
Farmerless
Farm

I

Sagebrush and lizards rattle and whisper behind me. I
stand in the moonlight, the hot desert to my back. It's
tomato harvest time, 3 A.M. The moon is nearly full and near
to setting. Before me stretches the first lush tomato field to be
taken this harvest. The field lies three hours northeast of Los
Angeles in the middle of the bleak silvery drylands of California's San Joaquin Valley. Seven hundred sixty-six acres,
more than a mile square of tomatoes — a shaggy vegetable
green rug dappled with murky red dots, 105,708,000 ripe
tomatoes lurking in the night. The field is large and absolutely
level. It would take an hour and a half to walk around it. Yet,
when I raise my eyes past the field to the far vaster valley floor,
and to the mountains that loom further out, the harvest is
lost in a big flat world.

This harvest happens nearly without people. A hundred
million tomatoes grown up, irrigated, fed, sprayed, now taken,
soon to be cooled, squashed, boiled, barreled and held at ready,
then canned, shipped, sold, bought, and after being sold and
bought a few more times, uncanned and dumped on pizza.
And such is the magnitude of the vista, and the dearth of human

presence that it is easy to look elsewhere and put this routine thing out of mind. But that quality — of blandness overlaying a wondrous integration of technology, finances, personnel, and business systems — seems to be just what the "future" that has befallen us has in store.

Three large tractors steam up the road toward me, headlights glaring, towing three thin latticed towers. The tractors drag the towers into place around an assembly field, then hydraulic arms raise them to vertical. Searchlights atop the towers soon illuminate a large sandy workyard where equipment is gathering — fuel trucks, repair trucks, concession trucks, harvesters, tractor-trailers towing big open hoppers. Now small crews of Mexicans, sunburns tinted to light blue in the glare of the three searchlights, climb aboard the harvesters; shadowy drivers mount tractors and trucks. The night fills with the scent of diesel fumes and with the sounds of large engines running evenly.

The six harvesting machines drift across the gray-green tomato-leaf sea. After a time, the distant ones come to look like steamboats afloat far across a wide bay. The engine sounds are dispersed. A company foreman dashes past, tally sheets in hand. He stops nearby only long enough to deliver a one-liner. "We're knocking them out like Johnny-be-good," he says, punching the air slowly with his right fist. Then he runs off, laughing.

The nearest harvester draws steadily closer, yawing and moving in at about the speed of a slow amble, roaring as it comes. Up close, it looks like the aftermath of a collision between a grandstand and a San Francisco tramcar. It's two stories high, rolls on wheels that don't seem large enough, astraddle a wide row of jumbled and unstaked tomato vines. It is not streamlined. It resembles a Mars Lander. Gangways, catwalks, gates, conveyors, roofs and ladders are fastened on all over the lumbering rig. As it closes in, its front end snuffles up whole tomato plants as surely as a hungry pig loose in a farmer's garden. Its hind end excretes a steady stream of stems and rejects. Between the ingestion and the elimination, fourteen laborers face each other on long benches. They sit on either side of a conveyor that moves the new harvest rapidly past them. Their hands dart out and back as they sort through the red stream in front of them.

Watching them is like peering into the dining car of a passing train whose guests are absorbed in an unchanging scene. The folks aboard, though, are not dining but are working hard for low wages, culling out what is not quite fit for pizza sauce — the "greens," "molds," "mechanicals," and the odd tomato-sized clod of dirt that has gotten past the shakers and screens that have already removed tomato from vine and dumped the harvest onto the conveyor.

The absorbing nature of the work is according to plan. The workers aboard this tiny outpost of a tomato sauce factory are attempting to accomplish a chore at which they cannot possibly succeed, one designed in the near past by some anonymous practitioner of the new craft of *management*. Later in the night these tomatoes are to be delivered to a cannery. A half-full tractor-trailer runs along next to the harvester, receiving its steady flume of tomatoes. Full trucks pull away; empty ones slide in next to moving harvesters. As per cannery contract, each of the semi-trailer loads of tomatoes must contain no more than 4 percent green tomatoes, 3 percent tomatoes suffering mechanical damage from the harvester, 1 percent tomatoes that have begun to mold, and 0.5 percent clods of dirt.

"The whole idea of this thing," a harvest executive had explained earlier in the day, "is to get as many tons as you can per hour. Now, the people culling on the machines strive to sort everything that's defective. But to us, that's as bad as them picking out too little. We're getting forty to forty-seven dollars a ton for tomatoes — a bad price this year — and each truckload is fifty thousand pounds, twenty-five tons, eleven hundred bucks a load. If we're allowed seven or eight percent defective tomatoes and we don't have seven or eight percent defective tomatoes in the load, we've given away money. And what's worse, we're paying these guys to make the load too good. It's a double loss. Still, you can't say to your guys, 'Hey, leave four percent greens and one percent molds when you sort the tomatoes on that belt.' It's impossible. On most jobs you strive for perfection. They do. But you want to stop them just the right amount short of perfection — because the cannery will penalize you if your load goes over spec. So what you do is run the belt too fast, and sample the percentages in the output from each machine. If the load is too poor, we add another worker. If it's too good, we send someone home."

The workers converse as they ride the machine toward the edge of the desert. Their lips move in an exaggerated manner, but they don't shout. The few workers still needed at harvest time have learned not to fight the machine. They speak under, rather than over, the din of the harvest. They chat, and their hands stay constantly in fast motion.

Until a few years ago, it took a crew of perhaps six hundred laborers to harvest a crop this size. The six machines want about a hundred workers tonight — a hundred workers for a hundred million tomatoes, a million tomatoes per worker in the course of the month it will take to clear the field. The trucks come and go. The harvesters sweep back and forth across the field slowly. Now one stands still in midfield. A big service truck of the sort that tends jet planes bumps across the field toward it, dome light flashing. Whatever breaks can be fixed here.

After the first survey, there is nothing new to see. It will be just like this for the entire month. Like so many scenes in the new agriculture, the essence of this technological miracle is its productivity, and that is reflected in the very uneventfulness of the event. The miracle is permeated with the air of everydayness. Each detail must have persons behind it — the inventions and techniques signal insights into systems, corporate decisions, labor meetings, contracts, phone calls, handshakes, hidden skills, management guidelines. Yet it is smooth-skinned.

Almost nothing anyone does here now requires manual skills or craft beyond the ability to drive and follow orders. And everyone — top to bottom — has got his orders.

The workaday mood leaves the gentleman standing next to me by the edge of the field in good humor. We'll call him Johnny Riley, and at this harvest time he is still a well-placed official on this farm. He is fiftyish and has a neatly trimmed black beard. His eyebrows and eyelashes match the beard, and his whole face, round, ruddy, and boyish, beams behind heavy black-framed glasses. He's a gladhander, a toucher, with double-knit everything, a winning smile that demands acknowledgment, and praise to give out. It is enjoyable to talk with him.

"There are too many people out here on the job with their meters running. We can't afford trouble with tomato prices so low. If something hasn't been planned right, and it costs us extra money to get it straightened out, it's my ass," he says.

The "us" on whose behalf Johnny Riley worries is called Tejon Agricultural Partners, Inc. But there are other eligible us's who would be displeased were there to be trouble with the harvest — a host of bodies, alive and corporate, with a stake in the success of the harvest: parent companies, general partners, limited partners, associated management corporations, processors sharing contractually the risks and profits of growing this crop. Tejon Agricultural Partners, in the immediate and legal sense, owns the tomatoes.

Tejon Agricultural Partners (TAP from now on) is one of those modern bodies that exists fully, in the eyes of the shareholders, the courts and the IRS, but is an elusive creature nonetheless — not at all the sort of thing one is used to regarding as a "farmer." That linguistic awkwardness is obviously felt locally, and the word of choice used by participants to describe such entities that cause food to grow in the Central Valley is *grower*. It applies to persons and corporations interchangeably.

The tomato harvester that has been closing for some time, bearing down on our outpost by the edge of the field, now is dangerously near. Behind the monster stretches a mile-and-a-quarter-long row of uprooted stubble, shredded leaves, piles of dirt and smashed tomatoes. Still Johnny Riley holds his ground. He has to raise his voice to make himself heard.

"I don't like to blow my own horn," he shouts, "but there are secrets to agriculture you just have to find out for yourselves. Here's one case in point. It may seem small to you at first, but profits come from doing the small things right. And one of the things I've found over the years is that a long row is better. Here's why. When you get to the end of a row, the machine" — Riley gestures up at the harvester, notices our plight, and obligingly leads me to one side. He continues, ". . . the machine here has to turn around before it can go back the other way. And that's when people get off and smoke. Long rows keep them on the job more minutes per hour. You've got less turns with long rows, and the people don't notice this. Especially at night, with lights on, row length is an important tool for people management. Three fourths of growers don't realize that. I shouldn't tell you so; it sounds like I'm patting myself on the back, but they don't."

And sure enough, as the harvester climbs off the edge of the tomato field and commences its turn on the sandy work road, the crew members descend from the catwalk, scramble to the ground and light up cigarettes. Johnny Riley nods knowingly to me, then nods again as a young fellow in a John Deere cap drives out of the moonlight in a yellow pickup to join us in the circle of floodlight the harvester has brought with it. It's as if he arrived to meet the harvester — which, it turns out, is what he did do. He is introduced as Buck Klein. Riley seems avuncular and proud as he talks about him. "I'm proud of this boy. Just a few years ago he was delivering material for a fertilizer company. Soon he was their dispatcher, then took orders. He organized the job. He came here to do pesticides, and we've been moving him up." Buck Klein keeps a neutral face for the length of this history, for which I admire him. He is of average height, sturdily built, sports a brush mustache that matches his short dark blond hair. He wears a western shirt, a belt with a huge buckle that says "Cotton" on it, and cowboy boots. He's come on business.

"We just got a truck back," he says, "all the way from the cannery at Fullerton — three hundred miles of travel and it's back with an unacceptable load. It's got twelve percent mechanical damages, so something's beating on the tomatoes. And this is the machine that's been doing it."

Johnny Riley appears to think for a moment. "We had three loads like that today. Seven, eleven and seventeen percent mechanicals. You got to take the truck back, get some workers to take out the center of the load and put in some real good tomatoes before you send it back. It ties up workers, and it ties up a truck. It's bad enough the trucks have to wait there to unload at the cannery without being sent back full."

"I'm going to go along on the harvester this trip," says Buck. "See if I can spot the trouble."

"Keep your eye on the speed," Johnny Riley says to Buck. To me he explains, "We pay a five-buck-a-load premium to the tractor driver on this machine to get him in a hurry to keep the machine on the field. Sometimes they get real hungry, and speed along a little. It lowers the quality because the machine is already set up for a certain speed. The incentive is a good management tool, but you got to watch the drivers. I took a psychology course once — as a management tool it was useful. In understanding the less educated people and why they get upset or do certain things, it's useful."

Buck eyes Johnny Riley, then walks past him to the harvester. The workers are climbing back aboard. We follow them. We look down at Johnny Riley, who is now two stories below on the ground. I cling to the starboard catwalk of the harvester. It humps off the dirt road and straddles the next unclaimed row of tomatoes. Buck rides next to me. I see Johnny Riley drive off in his car. Buck watches carefully as the first culls reach the lines of Mexican workers. Their hands flick out and back constantly. They don't look at us, and they no longer talk. Their arm motions pantomime dog paddling performed by swimmers clutching damaged tomatoes.

"It's not the tractor driver," Buck says finally. "You can see for yourself there are still too many mechanicals on the end of the belt, and they're doing everything right." He shoulders along behind the workers to the driver and shouts. The machine stops. Buck climbs down and disappears inside its innards. His cowboy boots stick out and up at an angle, all that's visible of him. When he finally wriggles back out, he says to the forewoman of the machine, "When you hit the other end of this row, stop and tell Gordon Fisher — he's the fellow with the crewcut who owns these pickers? — tell Mr. Fisher that the shaker rods here need adjusting. Understand?" The forelady nods and says, "Sí," and Buck thanks her and climbs down the ladder. We step off into the dark field. Soon the machine has left us behind. We walk back across the field toward the beginning of the row, where Buck's truck is still parked. Buck says he is beat. He's worked all day — the garlic harvest is on too. Then he went home and took his wife to Lamaze class — two months to the big day — and he napped for an hour, dozing next to his two-year-old boy.

"It's harvesting time," he explains. "Maturation goes on twenty-four hours a day. There's no overtime pay at my level — I'm management." He grins disbelievingly. "But farming work needs doing when it needs doing. You can't tell a plant, 'Stop growing, it's Sunday.' And you can't tell a bug, 'It's the Fourth of July, so stop eating the plants for the day.'" I comment to Buck that he sounds like a real farmer.

"I sort of pretend the ground's mine," he answers. "At least I know what I've put into the crops on it — I nursed these tomatoes start to finish. You talk to someone from another farm, you say, 'Over at our place . . . ,' so I guess you get to feel like it's yours. Of course I don't feel like I own it. I'd like to, but I don't have the money either. At times like this I'm pretty much in charge. Johnny Riley's probably on his way home to bed."

We walk through the tomato clutter for a while. Finally, Buck clears his throat and says, "O.K." It's as if he's decided he can go ahead and talk. "Don't get the idea it's all roses here," he continues. "Like these tomatoes. I sprayed them for worms last month — and they didn't need it. Worms are an urgent problem, and if you got them and don't spray for them, it's your butt. But we didn't really have them. I don't believe in wasting that money, if you know you don't have to. I was too young to carry weight — field got sprayed, and I had to sign the legal form as applicator, because I'm the one who took the pesticide course. I have been taught by good teachers. I was so frustrated I was ready to quit. I've seen so much money wasted on this place, things like that. Poor communications. It should be up to the field-level managers to spray or not. But I got a kid coming. After a while you just say, what the hell, I'm not going to take a reaming for sticking my neck out. You just go along with things."

As we get back to Buck's pickup truck, the two-way radio inside is squawking. Buck reaches in the window and answers it. A foreman named Les, across the night somewhere, is saying that there aren't enough trucks back now to keep up with the harvesters, and shouldn't he lay off one or two of the machines for the night?

"Where are the trucks?"

"Waiting in line at the canneries. Del Monte is about to go on strike — there's a slowdown. Also, that trailer with the leaking hopper, the driver won't take it on the road — doesn't want to get fined."

"Have Ken make him," says Buck.

"O.K.," says the voice. Buck checks to make sure Johnny Riley has really left the scene. He can't be raised on the radio. Buck calls the field office, gets some sleepy general's number, calls on a mobile telephone, and presents his case. He mumbles into the phone for a few minutes, then is back on the radio.

"Let's shut down one unit for a few hours, until more trucks show," he says. His rear is covered.

"We have budget sheets for every crop, and you better bend over if you stray very far from what it says you're going to do. It's what the management spends too much time worrying about, instead of how to make the crops better. It's all high finance. It makes sense, if you think about what they have in it. But I'll tell you something. It's expensive to farm here."

Buck points across the darkness, to the lights of the assembly yard. "Just beyond those lights there's a guy owns a piece — a section of land, and he grows tomatoes there too. A guy who works with the harvesters here, he knows tomatoes pretty well. And he says that guy has a break-even of about eighteen tons — eighteen tons of forty-dollar tomatoes pay his costs, and he's watching every row, growing better than thirty tons to the acre. Our break-even is twenty-four tons.

"Why? Because we're so much bigger. They give me more acres than I can watch that closely. They charge thirty-five or fifty bucks an acre management fee, good prices for this and that in the budget. And there is a stack of management people here, where that guy drives his own tractor while he thinks about what to do next. You can't beat him. This is not simple enough here.

"Here, they're so big, and yet they are always looking for a way to cut a dollar out of your budget. Trying to get more and more efficient. It's the workers who they see as the big expense here. They say, O.K., management is us, but maybe we can cut out some of those people on the harvesting machines. We can rent these machines from the custom harvester company for six dollars a ton bare. We got to pay the workers by the hour even when we're holding up the picking. Twenty workers to a machine, some nights, and two-ninety a worker is fifty-eight bucks for an hour of down time. You keep moving or send people home.

"Of course this will all be a thing of the past soon. There's a new machine out — Blackwelder makes it — and it's not an experimental model, I mean, it's on the job, at a hundred and four thousand dollars a shot, and it still pays. It does the same work, only better, with only two workers on it. It's faster, and there's no labor bill. It's an electronic sort. It has a blue belt, and little fingers and electric eyes, and when it spots a tomato that isn't right, the little fingers push it out of the way. You just set the amount of greens you want to be left alone, and it does that too. We're going to have two of them running later in the harvest, soon as they finish another job."

"What about the workers who have always followed the tomato harvest?" I ask.

"They're in trouble," says Buck, shaking his head. "They'll still be needed, but only toward the end of the harvest. At the beginning, most of what these cullers take away is greens. The electric eye can do that. But at the end of the harvest, most of what they take away is spoiled red, stuff that gets overripe before we pick it, and they say the machines don't do that as well. That leaves a lot of workers on welfare, or whatever they can get, hanging around waiting for the little bit we need them. They get upset about being sent away. This one guy trying to get his sister on a machine, he's been coming up to me all evening saying things about the other workers. I just ignore it, though. It's all part of the job, I guess."

The trouble in which California farm labor finds itself is old trouble. And yet, just a few years ago, when harvesting of cannery tomatoes was still done by hand, ten times the labor was required, to handle a harvest that yielded only a third of what Tejon Agricultural Partners and other growers expect these days. The transformation of the tomato industry has happened in the course of about twenty years.

Much has been written recently about this transition, and with good reason. The change has been dramatic and is extreme. Tomatoes we remember from the past tasted rich, delicate and juicy. Tomatoes hauled home in today's grocery bag taste bland, tough, and dry. The new taste is the taste of modern agriculture; it's the first thing new acquaintances comment about to writers involved with agriculture. Bev Bennett, writing for the Chicago *Sun-Times* remarked, "The American tomato has become a symbol of anything (and everything) that's wrong with our way of mass-producing food." She interviewed a spokesman for something called the California Fresh Market Tomato Advisory Board. "It's a matter of making the consumers more satisfied with the tomatoes they're buying," said the spokesman, as if what's missing is information rather than flavor. The Tomato Advisory Board's propagandistic mission seems difficult, but nevertheless simpler than would be the job of undoing the destruction of the good tomato. The ruination of the tomato was a complex procedure; the changes from what used to happen have occurred not merely on the farm but also in a number of other areas of human endeavor. Development of the modern tomato required cooperation from financial, engineering, marketing, scientific and agricultural parties that used to go their separate ways more and cross paths with less intention. These changes (their story *is* a good metaphor of what is happening to food production in general) have shifted to larger institutions' control of the money consumers spend on tomatoes. It is no more possible to isolate a "cause" for these shifts than it is possible to claim that it's the spark plugs that cause a car to run. However, we can at least peer at the intricate machinery that has taken away our tasty tomatoes and given us expensive, pale, scientific fruit.

Let us start then, somewhat arbitrarily, with processors of tomatoes — especially with the four canners, Del Monte, Heinz, Campbell, and Libby, McNeill & Libby, that sell 72 percent of the nation's tomato sauce. What has happened to the quality of tomatoes in general follows from developments in the cannery tomato trade. The increasingly integrated processors have consolidated, shifted, and "reconceptualized" their plants. The cheapest time for the processors to buy a perishable harvest, of course, is at harvest time. Because the few manufacturers make most of the wide variety of tomato products, it is to their advantage to buy tomatoes then, but to wait as long as possible before committing themselves to any particular tomato product. In the fast world of marketing processed tomatoes, the last thing executives want is to be caught with too many cans of pizza sauce, fancy grade, when the marketplace is starved for commercial catsup.

What processors do nowadays is capture the tomatoes and process them until they are clean and dead, but still near enough to the head of the assembly line so they have not yet gone past the squeezer that issues tomato juice nor the sluice gate leading to the spaghetti sauce vat or the paste vat or the aspic tank, or to the cauldrons of anything in particular. The mashed stuff of tomato products is then stored until demand is clear. Then it's processed the rest of the way. The new manufacturing concept is known in the trade as aseptic barreling, and it leads to success by means of procrastination.

The growers supplying the raw materials for these tightly controlled processors have contracted in advance of planting season for the sale of their crops. It's the only way to get in. At the same time, perhaps stimulated by this new guaranteed marketplace — or perhaps stimulating it — these surviving growers of tomatoes have greatly expanded the size of their plantings. This interaction of large growers and large processors has crowded many smaller growers out of the marketplace, not because they can't grow tomatoes at least as cheaply as the big growers (they can) but because they can't provide large enough units of production to attract favorable contracts with one of the few canners in any one area.

In turn, the increasing size of tomato growing operations has encouraged and been encouraged by a number of developments of technology. Harvesters (which may in fact have been the "cause" precipitating the other changes in the system) have in large part replaced persons in the field. But these new machines became practical only after the development of other technological components — especially new varieties of tomato bred for machine harvesting, and new chemicals that make machine harvesting economical.

Wild progenitors of *Lycopersicon esculentum* evolved (according to a recent cover article in *Scientific American*) around the Andes, and were domesticated by Mexican Indians. The fruit is called *Tomatl*, from which "tomato" derives, in the Indian language of Nahua. *Pomi d'oro* showed up in Italian herbals in the mid–fifteen hundreds, listed chiefly as a medicine related to belladonna and mandrake. The plant found its way to North America by sea, reimported by early European settlers. In colonial times it was not uncommon. Thomas Jefferson grew tomatoes.

The story that schoolchildren are told about tomatoes having once been thought poisonous apparently has some basis in fact. An alkaloid called tomatine, which is slightly toxic, lurks in leaves and in green tomatoes but turns into something else in ripening fruit. It may have given a few ancient diners stomachaches. But those early herbalists saw people eating tomatoes, and Jefferson records tomato eating as a matter of course.

What is remarkable about the tomato from the grower's point of view is its rapid increase in popularity. In 1920, Americans each ate 18.1 pounds of tomato. These days we each eat 50.5 pounds of tomato. Half a million acres of cropland grow tomatoes, yielding nearly 9,000,000 tons, worth over $900,000,-000 on the market. Today's California tomato acre yields 24 tons, while the same acre in 1960 yielded 17 tons and in 1940, 8 tons.

The increased consumption of tomatoes reflects more generally changing eating habits. Most food we eat nowadays was prepared, at least in part, somewhere other than in the home kitchen, and most of the increased demand for tomatoes is for processed products: catsup, sauce, juice, canned tomatoes and paste for "homemade" sauce. In the nineteen twenties tomatoes were grown and canned commercially from coast to coast. Small canneries persisted into the nineteen fifties. Tomatoes were then a labor-intensive crop, requiring planting, transplanting, staking, pruning. And, important in the tale of changing tomato technology, because tomatoes used to ripen a few at a time, each field required three or four forays by harvesting crews, which recovered successively ripening fruits.

The forces that have changed the very nature of tomato-related genetics, farming practices, labor requirements, business configurations, and buying patterns start with the necessity, built so deeply into the structure of our economic system, for the constant perfection of capital utilization. The details of the change have recently been documented by numerous journalists. Jim Hightower's *Hard Tomatoes, Hard Times* traced the use of public money to do research on tomato harvesting machines, leading to great private profit and public damage. William Friedland and Amy Barton's *Destalking the Wily Tomato* also traced the history and impact of the transformation.

Some researchers sometimes seem to imply, however, that the new mechanization is a conspiracy fostered by fat cats up top to make their own lives softer. A more accurate assessment of the transformation of the tomato industry would be easier on individuals, but even harder on "the system." Although there are surely greedy conspirators mixed in with the regular folks running tomato farms and tomato factories and tomato research facilities, the impulse for change at each stage of the tomato transformation — from the point of view of those effecting the change — is merely pressure *to meet the competition.* If at some time between the late fifties, when it all started, and the present day, when the slums of California are crowded with displaced *braceros* and the hills of Ohio with displaced processing plant workers, some archangel of investigative reporters had fallen out of the clouds to interview the farmers, corporate executives, professors of agricultural engineering, bankers, planners of larger canneries working for multinational conglomerates, geneticists trotting about fields of increasingly tough and tasteless prototype tomatoes, and had that repertorial archangel asked each of these partners in the labor of tomato transformation why they participated in this activity, each would have been able to answer, "To keep my job, to keep my farm, to have success in getting grants, to protect my position in the corporation, to keep abreast of what similar businesses are doing." And then each respondent would have been able to drive to one of those low, new California churches, pray with a clear conscience, give to the United Fund, the University Alumni Fund, give blood to the Red Cross, and remain at peace, without questions about the nature of progress.

Even in the nineteen twenties more tomatoes were grown commercially for processing than for fresh consumption, by a ratio then of about two to one. Today the ratio has increased to about seven to one. Fifty years ago, California accounted for about an eighth of all tomatoes grown in America. Today, California grows about 85 percent of tomatoes. Yet as recently as fifteen years ago, California grew only about half the tomato crop. And fifteen years ago, the mechanical harvester first began to show up in the fields of the larger farms. Its availability transformed both the habits and economics of tomato farmers.

Before the harvester came, the average California planting was of about 45 acres. Today, plantings exceed 350 acres. Tomato production in California used to be centered on family farms around Merced. It has now shifted to the corporate farms of Kern County, where TAP operates. Of the state's 4,000 or so growers harvesting canning tomatoes in the late sixties, 85 percent have left the business since the mechanical harvester came around. Estimates of the number of part-time picking jobs lost go as high as 35,000.

The introduction of the harvester brought about other changes, too. Processors thought that tomatoes ought to have more solid material, ought to be less acid, ought to be smaller in size. Engineers called for tomatoes that had tougher shells and were oblong so they wouldn't climb up and roll back down tilted conveyor belts. Larger growers, more able to substitute capital for labor, wanted more tonnage per acre, resistance to cracking from sudden growth spurts that follow irrigation, leaf shade for the fruit, to prevent scalding by the hot sun, determinate plant varieties that grow only so high, to keep those vines in rows, out of the flood irrigation ditches.

As geneticists selectively bred for these characteristics, they lost control of others. They bred for thick-walledness, less acidity, more uniform ripening, oblongness, leafiness, and high yield — and they could not also select for flavor. And while the geneticists worked on tomato characteristics, chemists were perfecting an aid of their own. Called ethylene, it is in fact also manufactured by tomato plants themselves. All in good time, it promotes reddening. Sprayed on a field of tomatoes that has reached a certain stage of maturity (about 15 percent of the field's tomatoes must have started to "jell"), the substance causes the plants to start enzyme activity that induces redness. About half of the time a tomato spends between blossom and ripeness is spent at full size, merely growing red. (The stages of this ripening are called, in the trade, immature greens, mature greens, breakers, turnings, pinks, light reds, and reds.) Ethylene cuts this reddening time by a week or more. It clears the field for its next use. It recovers investment sooner. Still more important, it complements the genetic work, producing plants with a determined and common ripening time so machines can harvest in a single pass. It assures growers precision. On schedule, eight or ten or fourteen days after planes spray, the crop will be red and ready. The gas complements the work of the engineers too, loosening the heretofore stubborn attachment of fruit and stem. It allows the tomato to be easily shaken free of its mooring.

The result of this integrated system of tomato seed and tomato chemicals and tomato hardware and tomato know-how has been, of course, the reformation of tomato business.

According to a publication of the California Agrarian Action

Project, a reform-oriented research group located at Davis (whose findings are reflected throughout this section), the effects of an emerging "low-grade oligopoly" in tomato processing are discoverable. Because of labor savings and increased efficiency of machine harvesting, the retail price of canned tomatoes should have dropped in the five years after the machines came into the field. Instead, it climbed 111 percent, and did so in a period that saw the price of all processed fruits and vegetables climb only 76 percent.

There are "social costs" to the reorganization of the tomato processing industry as well. The concentration of plants concentrates work opportunities formerly not only more plentiful but more dispersed in rural areas. It concentrates herbicide, pesticide and salinity pollution problems.

As the new age of cannery tomato production has overpowered earlier systems of production, a kind of flexibility in tomato growing has been lost that once worked strongly to the consumers' advantage. The new-style, high-technology tomato system involves substantial investment "up front." New-style seed, herbicides and pesticides, machinery, water, labor, and "management" of the crop are all costly, involving great risk. Everything must be scheduled far in advance.

In order to reduce the enormous risks that might, in the old system, have fallen to single parties, today's tomato business calls for "jointing" of the tomatoes. Growers nowadays share the risks of planting, raising, harvesting and marketing — by farming together with a "joint contractor." The tomatoes grown by Johnny Riley and Buck Klein on land held by Tejon Agricultural Partners, general and limited, were grown under a joint contract with Basic Vegetable Products, Inc., of Vacaville, California. TAP's president at the time, Jack Morgan, was previously executive vice-president of Basic Vegetables.

Because the formulation and execution of contracts under which farm products are jointly grown these days are costly and various, and require raising of risk capital and the supervising of its expenditure, such deals are expensive both to set up and administer. Processors and distributors find it cheapest to contract with a few large growers rather than with many smaller growers. The tomato-growing business situation is becoming so Byzantine and costly that the "per-unit cost of production," the cost to a grower of producing a pound of tomatoes, no longer is the sole determinant of who gets to grow America's tomatoes. Once, whoever could sell the cheapest won the competitive race to market. Today, the cost of doing all business supersedes, for large-scale operations, simple notions such as growing tomatoes cheaply. Market muscle, tax advantages, clout with financiers, control of supply — all affect the competitive position of TAP as much as does the cost of growing tomatoes.

The consequence of joint contracting, for the consumer, is a slightly higher priced tomato. Risks that until recently were undertaken by growers and processors and distributors separately, because they were adversaries, are passed on to consumers now by participants that have allied. Growers are more certain they will recover cost of production.

Howard Leach, who was president of TAP's parent company, Tejon Ranch, at the time of the tomato harvest, understood very well the economic implications for consumers of joint contracting.

"Productivity lessens," Leach explained to me. "Risk to the producer lessens, which is why we do it. The consumer gets more cost because the processor who puts money in will try to lower supply until it matches the anticipated demand. If you're Hunt-Wesson, you gear up to supply what you forecast that sales will be. You want an assured crop, so you contract for an agreed price. You're locked in, and so is the farming organization. But they are locked into a price they are assured of, and they are big enough to affect the supply."

Under this sort of business condition, the marketplace is fully occupied by giants. It is no place for the little guy with a truckload or two of tomatoes — even if his price is right. Farmers who once planted twenty or thirty acres of cannery tomatoes as a speculative complement to other farming endeavors are for the most part out of the picture, with no place to market their crops, and no place to finance their operating expenses. As John Wood, a family farmer turned corporate manager, who currently runs TAP, puts it, "The key thing today is the ability to muscle into the marketplace. These days, it's a vicious fight to do so." And Ray Peterson, the economist who was vice-president of Tejon Ranch when Howard Leach was its president, sums up the heightened importance of the business side of farming now that the new technology has increased the risk and scale of each venture. "Today," he says, "vegetable farming is more marketing than farming."

The "jointing" of vegetable crops integrates the farming operation with the marketing, processing and vending operations so closely that it takes teams of lawyers to describe just where the one leaves off and the other begins. And joint contracting is only one of several sorts of financial and managerial integration with suppliers and marketers that occur in the new tomato scene. Today chemical companies consult as technical experts with farming organizations. Equipment companies consult with farming organizations about what machine will do the jobs that need doing. Operations lease equipment from leasing companies run by banks that also lend them funds to operate. Financial organizations that lend growers vast sums of capital for both development and operations receive in return not merely interest but negotiated rights to oversee some decision-making processes. Agricultural academics sit on agribusiness corporate boards.

Today the cannery tomato farmer has all but ceased to exist as a discrete and identifiable being. The organizations and structures that do what farmers once did operate as part and parcel of an economy functioning at a nearly incomprehensible level of integration. So much for the tasty tomato.

II

TAP was designed and its birth attended by a parent company, Tejon Ranch Corporation. The ranch, which is listed on the American Stock Exchange, has tangible assets of considerable magnitude. Foremost of these is a chunk of land — about three hundred thousand acres, located mostly in Kern County just above Los Angeles. Tejon Ranch also owns a seed company, cattle grazing and feedlot operations, oil reserves, a cement plant site, and the land around five consecutive freeway interchanges. The large piece of land — three hundred thousand acres is the size of two Manhattan Islands with a neighborhood of the Bronx thrown in for elbowroom — follows in part the boundaries of an original Spanish land grant. It was acquired for speculation early in this century by a group of Los Angeles businessmen, chief among them members of the Chandler family, which today owns the Los Angeles *Times*, and controls the Times-Mirror Syndicate, a company about one hundred fiftieth in rank on the *Fortune* magazine list of the largest companies in the country. Chandler family members still own large portions of Tejon Ranch Company stock, as do the Times-Mirror Syndicate and the Superior Oil Company.

The investors set out to run a cattle grazing business but found that their land, located south of Bakersfield oil fields, itself covered a lake of oil. Many hundreds of wells still pump oil on Tejon Ranch land, and other wells there stand capped against rising oil prices and improved dry-well technology. Some of the ranch's oil wells are on the twenty-one thousand acres farmed by TAP, but they are owned by the parent company and leased to others. TAP employees farm around them.

TAP was first conceived in response to some urgent news from the state government in 1971. The California Water Project was to bring irrigation water — a great concrete-walled river of water, an aqueduct from a thousand miles to the northeast, a ribbon of canals, pipeline, sluiceways passing over highways — right through Kern County, through the section of the ranch's land now farmed by TAP. The state's water project offered a "now or never" deal — sign up immediately for the water you will want, or miss the opportunity forever. The water would turn sere grazing dryland into some of the most fertile and, what's as important in the new agriculture, some of the most predictable farmland in the nation. The valley's few inches of natural rainfall a year, most of which falls in winter, allow uncommonly unvarying and therefore controllable crop-growing conditions. It would be a windfall for anybody capable of rallying the stupendous start-up costs that irrigation entails.

To shift from the low-keyed land supervision practiced as a sideline by the ranch before the water project came, to the capital-intensive enterprise suggested by the high costs of irrigated agriculture involved a major corporate policy decision.

Tejon Ranch Corporation higher-ups thought about this opportunity and were not at all sure it was a windfall. They nevertheless felt obliged to sign on while the circus was in town. The "now-or-never" form of the state's offer encouraged corporations to get into intensive farming. Because of the development costs — made even higher by high projected fees for water — the offer pushed participants into the development of permanent crops —

vineyards, nut orchards, citrus and fig groves — because they would eventually yield high per-acre returns on investment. As agricultural economist Ray Peterson remembered the time of decision, "It was basically a one-time-only offer. If you wanted the water, you had to sign up for it by a certain date, and you had to pay the fixed costs of bringing it out after that, whether you used it or not. It amounted to the government promoting large-scale agriculture. You couldn't feel your way in, and it was no place for the small operator."

In order to use the water they decided to purchase, the new farming enterprise would have to gather and spend many tens of millions of dollars. They would have to spend money to survey and level roads and vast fields, lay irrigation pipeline, excavate tailwater ditches, install in-the-ground permanent sprinkler and drip irrigation outlets, construct pumping stations, work buildings, offices, and housing, and acquire a lotful of farming equipment. They would have to develop and plant and tend square miles of crops at costs of thousands of dollars per acre. In addition, they would have to hire away from elsewhere scores of personnel, from ditchers, tractor drivers, mechanics and agronomists to corporate heads, lawyers, accountants and engineers. They would have to formulate a business structure to handle this development. And they would require financing with both loans and investment capital. A new farm, of course, has no earned cash at all until it has set up and has had a season to plant, reap and market. "Permanent crops" take four to eight years before they mature enough to begin to return on investment.

The plan to establish such permanent crops virtually from scratch on twenty-one thousand acres of ground (an area the size of a large New England town) would cause capital to evaporate as quickly as irrigation water in the desert sun, and keep it evaporating for some years. The intention to undertake these development losses constituted something of an asset in itself. It allowed Tejon Ranch's executives to craft an enterprise that would attract investors' capital by offering to give them, in return, credit for some of the development losses — they could then claim them as deductions against taxes due on their other sources of income.

Tejon's executives set up the new corporation, Tejon Agricultural Partners (eventual owners of the tomatoes being mechanically harvested in the desert night), with such benefits in mind. TAP was organized as a limited partnership — a company owned by partners both limited and "general" (limited partners share, only to the extent of their investments, in the risks of the company; a general partner risks all its assets, and it retains all of the power to control the operations of the partnership).

In the case of TAP, the general partner was not a person, but yet *another* corporate body, called Tejon Agricultural Corporation, also set up for the occasion by Tejon Ranch and wholly owned by them. Tejon Ranch transferred to the general partner, Tejon Agricultural Corporation, 21,000 acres of its land, appraised for them at about $13.7 million (it had previously been valued at $125,000) and also turned over to them about $3 million for use in the operation. The general partner thus, according to themselves, brought into the deal about $17 million worth of value in land and money.

The prospectus of Tejon Agricultural Partners was issued on July 19, 1972, and offered to do business only with sophisticated investors — those able to understand the risky nature of the proposition and who were either in the 50 percent tax bracket or held at least $200,000 worth of other assets. The prospectus offered $16 million worth of "units" for sale at $1,000 apiece, and required that each limited partner purchase 5 or more of them. Each unit bought would be subject to an additional assessment of up to $200 if the partnership required further funding, which it eventually did.

In 1972 (when this business arrangement was made) even more than today, high-income taxpayers were actively searching for "tax writeoff" investments. Laws of the time, since revised, made such investments very attractive. TAP's structure was tailored to attract capital in such a marketplace. The plan was formed at the tail end of an era in which it was still possible for an organizer with a fresh idea — even in the absence of assets or corporate history — to write up a prospectus and raise huge sums of money by promising investors that a partnership would produce writeoffs against their other tax payments. The deals were structured to allow writeoffs of more money than investors actually placed at risk in a scheme. When one of these plans worked out as proposed, wealthy investors enjoyed tax benefits that saved them, in the very first year, more than they'd risked — sometimes two and three times over.

Developed in the shadow of a highly respectable parent company and possessing land, some funds, and credit, TAP seemed far sounder than most of the plans that had succeeded. It was possibly the last major agricultural project funded before reform "tax loss farming" legislation in the early 1970s closed most of the gaps still surviving earlier reform legislation.

In theory, tax laws of the sort TAP's designers had in mind encouraged kinds of high risk and capital-intensive businesses that legislators found would not otherwise be started, but that would serve in the public interest. Such "tax loss schemes" are usually explained by their defenders as "deferring," not eliminating, tax payments. The critics, whose opinions eventually helped cause those legislative reforms, argued that tax-loss farming schemes amounted to enforced government subsidization of corporate competition with family farmers.

At any rate, TAP fitted existing laws in a way that held promise of benefiting its partners considerably. Within three months of the initial offering, approximately thirteen hundred buyers, described to me by one Tejon Ranch official jocularly as "one thousand three hundred dentists," had bought up all sixteen million dollars' worth of the units offered. Investors included not only dentists, but doctors, persons of leisure, at least one internationally famous sculptor, and the American Express Company.

The twenty-one thousand acres to be farmed were not simply turned over to the new business forever. The land was encumbered — a mortgage against it (a "deed of trust" in California) was signed over to John Hancock Life Insurance Company, of Boston, Massachusetts. Secured by this collateral, and also by an overriding interest in the developing crops, Hancock made a loan to the partnership amounting to about $27,000,000, at an interest rate of about 8½ percent. Built into the prospectus, and therefore consented to by the partners at their time of purchase, was an agreement that TAP make an annual payment to Tejon Agricultural Corporation (the general partner, wholly owned by the parent company, Tejon Ranch) of a land use fee of about $1.1 million, which is equal to 8½ percent interest on the $13.7 million appraised value of the 21,000 acres.

Tejon Ranch, which in a sense enjoyed this "land use fee" paid to its wholly owned subsidiary (it is known as the "Guaranteed Payments to General Partner on Land Contribution"), also occupied the position described in the prospectus as "farm manager." In return for this, it would receive "farm management fees" of $50 an acre during the first year of operation and $35 an acre in subsequent years — fees that would amount to about $1,000,000 in the first year, and $750,000 in subsequent years.

The limited partners, by the terms of the offering prospectus, were confined in their participation to a thirty-year term. In 1997, like Brigadoon the morning after, TAP would cease to exist, and all its assets would be turned back to the general partner, after creditors were paid off. The developed and producing vineyards, nut orchards, citrus groves, and cropland, all the fixtures, all the inventory, the prospectus read, would cease to be the property of TAP. Tejon Ranch would come to own a large and fully developed corporate farming operation, which limited partners, and America's taxpayers, helped to create.

"The dentists" who, to a significant degree, financed this operation that they would eventually not own, nevertheless bought shares with expectation of appreciable personal gain. As one limited partner recounted the motive for his purchase, "Farming is the backbone of the country. I'd had experience with other farming investments. And, yes, I had tax problems at the time. I expected that the tax benefits I would receive would help with those. But no tax shelter is good unless the business is successful. I had other information not in the prospectus — my broker showed it to me — that projected a return after twenty years, or sooner, of four or five to one, which seemed reasonable and fair for an investment of this type."

The limited partners bought "units" with the anticipation of two sorts of benefits. During the initial period of high start-up costs they could expect losses to write off against other income (the income that had made them eligible to buy in by placing them in the 50 percent tax bracket). And they hoped eventually to enjoy an increase in the value of their "units" — once TAP developed and its income from sale of crops began to pay back costs. They would, said the prospectus, be paid back for their initial contributions, plus 7 percent interest ("not compounded").

Then they would share with the general partner in further profits — but only after other parties were paid. Tejon Ranch was the next in line to reclaim its "cash contribution." Also, their "farm management fee" was subject to "escalation" every five years in relation to the Cost of Living Index. And an "incentive fee" equal to 25 percent of TAP's pretax profits would also be due to the "farm manager." After that, the limited partners would share what remained in the pot.

To go over the same ground with less detail, the business situation set up by the prospectus looked like this. Limited partners bought shares in a farming partnership, TAP. Their participation was limited to receiving tax credit for portions of the development costs exceeding their own investments, which credits they could use to offset their taxes on other income. In the long run, they might receive back the cost of their investment, and, later still, a share of the profits of the going concern. The general partner of TAP was another company, Tejon Agricultural Corporation, in whose hands Tejon Ranch, the parent company, had placed twenty-one thousand acres and three million dollars.

If the general partner simply were to wait until 1997, the whole kit and caboodle would be turned over to it, *exeunt limited partners.*

At the same time all this farming and tax losing and waiting-until-1997 was going on, the 21,000 acres under development also served as equity in a secured loan made by John Hancock Life Insurance (there are other loans as well) for $27,000,000. This part of the deal is analogous to a storeowner who takes on some partners while also securing a bank mortgage with the shop, all the time going right ahead doing business.

In the end, before permanent crops matured, before income began to exceed expenses — indeed, before development plans had been completely carried out — TAP ran low on money. It ran low on the eighteen-odd millions that the partners had eventually put in. It ran low on the twenty-seven millions that John Hancock Life Insurance had lent. It ran low on five millions more that CitiBank had lent it, and it ran low on the money that Tejon Ranch had put in as a cash contribution. Rumors flew transcontinentally in 1977 that TAP was contemplating bankruptcy.

The 1977 annual report of the parent company started with a notice "to our shareholders" from Calvin Walters, the new president of Tejon Ranch Company. "Operating cash deficits of TAP became critical during the year. . . . Because lack of operating capital threatened continuance of the farming program, in February, 1978 the Partnership's major lenders were advised that TAP did not have sufficient working capital to maintain the Partnership's plantings. . . . it is Management's opinion that the Ranch Company's contingent right to receive remaining Partnership property upon termination or dissolution of the Partnership, after satisfaction of creditors' claims, is of uncertain value. . . ." The notice to shareholders explained some reasons for the "cash crunch," and outlined a plan for dealing with the situation.

It seemed to be a fulfillment of the warning set off in boldface type, entirely capitalized, and centered on page one of the 1972 offering prospectus:

THIS OFFERING INVOLVES A HIGH DEGREE OF RISK AND SUBSTANTIAL PAYMENTS TO THE GENERAL PARTNER AND ITS AFFILIATE PRIOR TO ANY CASH DISTRIBUTION TO THE LIMITED PARTNERS.

What took place to bring about the fulfillment of this warning can best be understood by returning to the time and atmosphere of the enterprise when it was starting and functioning.

In 1972, having raised and borrowed its start-up capital, TAP's first order of the day was to reoccupy the land that for years had been leased out to grazers and wellhead irrigators by Tejon Ranch Corporation. Ray Peterson explained, "When the full water supply came, higher-income crops became possible. We use an acre foot-and-a-half of water just to grow wheat, and the water costs us thirty-five dollars and up for each acre foot. Vegetables take three acre feet. Kansas grows its wheat with free rain water. The costly water pushes you into the higher-value crops, and we wanted more control to see that our investment money is well spent. We gave the tenants a year's notice, and they moved on. A few were near retirement. In the cases with financial problems they were probably better off forced to give up their operations. The majority were farming only a portion of their operations on TAP land anyway, and they found other land. A good operator will make it anywhere.

"A deed means what it says; a lease also means what it says. I have human compassion. I like to see people doing well who are trying and working hard. To the extent that a mutually beneficial agreement is possible, good. But merely because of past history, there's no reason to let someone go on farming contrary to the good of this company. You have to do things that are unpleasant. That doesn't mean they're wrong."

Some of the lessees, who had farmed the land for long enough to consider it a habit and to feel proprietary about it, did of course feel wronged. Today one can drive for hours in Kern County, witness tractorwork, harvesting, cultivating on every section of land, and never once see work performed by any but employees.

Having moved the old guard out, TAP and its allied forces set about developing its new 21,000-acre food factory with vigor. They did everything first class. The limited partners were not about to criticize high development costs. The irrigation in many sections of the farm includes computer-operated permanent piping systems buried among the many square miles of tree crops.

Most frequently, the new enterprise leased rather than purchased equipment. They leased nearly a million dollars' worth of tractors, a quarter of a million dollars' worth of pickup trucks and staff cars, about three million dollars' worth of irrigation equipment, and more than a million dollars' worth of harvesting equipment.

In just two years' time, they cleared and tilled many thousands of acres of rough ground. They planted grapevines, whose long rows in some sections they kept true by laser beam in order to permit machine harvesting. They planted nut trees, fig trees, and citrus. They installed thousands of miles of irrigation pipe and built hundreds of miles of ditch and roadway. They also erected a three-acre greenhouse, in which grew virus-free grape cuttings to start the vineyard, and, later, potted plants for discount-store florist sections. They built mechanical shops, field offices, an executive enclave, reconditioned former lessees' homes to offer their own foremen, assembled (mostly by their leases) 66 tractors, (one tractor for each 420 acres of permanent crop or each 200 acres of row crop), 56 pickups and cars, 22 bigger trucks, 8 crawler tractors, thousands of fixtures to control the newly arriving water. They hired, jostled, proved and shuffled a large staff and hundreds of field workers. They commenced to farm.

By 1976 TAP had growing the following permanent but still maturing crops: 3,915 acres of almonds, 825 acres of figs, 7,769 acres of grapes, 531 acres of oranges, 1,064 acres of pistachios and 462 acres of walnuts. In addition they had, as promised in their prospectus, taken to improving their soil tilth as well as cash flow by growing 653 acres of cotton, 1,571 acres of wheat, 608 acres of onions, 285 acres of spring carrots and still more fall carrots, 304 acres of garlic, 190 acres of watermelons, and experimental acreages of broccoli, lettuce, and cauliflower. Much TAP acreage remained that hadn't been farmed in years and was slowly being claimed and leveled. Lessees still held 1,600 acres of land. And tomatoes were being harvested from 777 acres.

The coming of the nearly incomprehensible level of agricultural-business integration has created a new and sheltered enclave, a world of the few humans needed to attend the machinery, to keep it rolling and pointed right, working up to snuff, in the right place at the right time, to keep the genetically reshaped plants perked up and to keep the new chemical fogs billowing in as per schedule, to plan these events, to handle budget and contractual problems and plan next year's budget and contractual problems, to talk money, and pay and, most important, to collect money. The new world of high-tech farming has gathered up a small crew of "managers," who, as one described their activities, "interface between systems and persons."

Buck Klein, tomato harvest troubleshooter, is one such gentleman. Many of the new management class come to the valley farming scene from other — frequently southern — farming areas like construction workers wandering up to Alaska. Buck Klein, though, is a son of the valley. He's never gone far from home.

"I grew up right in Shafter, right outside of Bakersfield. You could walk across town if you wanted the job of walking across town. My grandmother lived across town, right on the edge. We'd drive across town to there, and we'd hunt."

It's noon, still in the midst of the tomato harvest. Walking with Buck into Denny's Restaurant, along Highway 5 — the "Golden State" — on Tejon Ranch's land and leased by the parent company to a concessionaire, a small crowd of men still talk about hunting. It's a procession of Marlboro men, men who possess the highest level of skill that yet leaves them working with hands as well as heads, men who can *do* things, men who are in between, neither carefree workers following orders and applying their skills, nor concerned executives with the autonomy to express their competence.

Buck favors Winchester rifles. He also likes Ford trucks. He tells the others about hunting rabbits when he was younger on the flats outside of Shafter, on the alkali lake-bottom owned by a poor Basque shepherd named Juan. Plenty of rabbits, but the interesting thing was Juan, who made a cagy deal when water was coming in a few years back and sold out his dollar-an-acre spread for a thousand an acre, then disappeared "off east somewhere to herd more sheep, probably."

The other men taking seats at the table are dressed like Buck — tight-fitting Levi's, shirts with patterns and pointy collars, cowboy boots and belts with large brand-name buckles. Today Buck's says "John Deere." The words are cast in aluminum, superimposed over a bas relief of a new mechanical cotton harvester — a machine with a recent history like that of the tomato harvester. (John Deere and International Harvester together share 57 percent of the U.S. tractor market.) The next

man to Buck's left has a buckle that says "Cotton" in raised
boldface letters, stamped in brass over a bas relief of cotton
bolls. (Kern County grows a quarter of a million acres of
cotton. It's the bread and butter of many of the growers — the
crop that gives reliable income to the medium-sized family-run
farms that still predominate right around Bakersfield.) The
next fellow sports a simple buckle — block letters that read
"MONSANTO" — no picture behind it. Next buckle, "De-
Kalb" over a long, gracefully swollen ear of corn. (DeKalb
AgResearch and Pioneer Hi-Breed International together supply more than half of all the hybrid seeds sold in America.)
Angelo Mazzei's belt buckle also says "John Deere" on it, laid
over the figure of a leaping fawn. Angelo says he collects
buckles, also the bill-brimmed hats that seem to be the universal and undoffed headgear of all occupants of Denny's
Restaurant save for the waitresses. The belts resemble those
stove-door-sized prize belts awarded professional wrestlers. They
are part of the regalia of a fantasy of strength that keeps com-
promised heroes in self-respect. A belt like that's a lot to take
off, a veritable ship's hatch guarding the valuable cargo below
from the nasty batterings of a cruel sea of farm management.

Buck is done talking about rabbit hunting, now turns his
thoughts to the problems of supervising the tomato harvest in
a situation where he feels the realities of jointly contracted
crops and multitiered management inhibit his managerial hand.

"I know how to grow good tomatoes. I've been taught by
people who knew all the tricks. But I say something needs
doing, and somebody else says something else, and pretty soon
somebody's on the phone to the office and somebody's on the
radio from the office, and all that crap from harvesting people
ruins the enjoyment of growing a good crop. Everything you
do has to be checked out. There's always a pretty good question about whether your boss will cover your ass or not."

The others at the table know Buck's dilemma and feel they
share it. Angelo Mazzei, the young industrial engineer hired
to start up and run TAP's huge maintenance shop, has charge

of the extensive inventory of equipment. It's a big-time operation — a staff of twelve fixes things twenty-four hours a day. They even invent new equipment. All other operations of TAP rent their machinery from Angelo's shop. It's set up that way for what Angelo calls "reasons of internal bookkeeping." He is quiet, attentive, self-possessed, and astonishingly courteous. He is not a josher. He is another son of the valley. His uncle was one of the more adventurous of the private vegetable farmers in the area, used to lease Tejon land. Angelo is angelically chubby, slightly bald, thirtyish, and wears sturdy chrome-rimmed glasses.

"I put in my time," Angelo says about his job. "I enjoy the harder mechanical problems. I enjoy the programs with the men in my department. We're having a safety contest now — giving away wristwatches, a thirty-dollar item that saves far more than that on insurance. It's when I have to reach outside my department that I don't enjoy my days here. You can't count on what people will do with what you tell them." Angelo nods over to the fellow sitting next to him, a hulk, a giant of a man dressed like a gangster in a pale yellow doubleknit suit and loud gold sportshirt. "Bill Fryer here will know what I'm talking about." Fryer, an accountant, once started up a chain of fried chicken restaurants (Chesley Fipps Fish and Chicks), then ran into "cash-flow problems." His salary at TAP tides the restaurants over until things improve. "I'm director of administration," he says. "It's a liaison job with the parent company in L.A. I handle all the bull — except production. I take fifty supervisors' complaints and detailed troubles about budgets. I also get it from the L.A. management."

"What is Angelo talking about that he says you know about too?"

"I don't want to talk to you. It won't do me any good at all. I like you, but I don't think it can do me any good, and it could hurt. I took this job because I need it."

"What Angelo's talking about," says a new voice, belonging to a barrel-chested man in a red and green Florida tourist shirt and silvered sunglasses, "is just how frustrating it is to try to do a good job. I make pretty good money. I know electronics, plumbing, water. I work with my men — I'll do any job they'll do — makes them respect you. But you get a good idea, and by the time someone says yes, it's too late to do it anymore. Or else they say yes, and you try it, it doesn't work, and they say, 'Who the hell told you to do that?' "

"Your rump!" says Buck.

Angelo nods assent.

Also sitting at the table, trading quips with the best of them, prepared to pick up everyone's tabs and waiting for their moment to do business, are two chemical-company salesmen. The salesmen know a secret, which they reveal by their attitudes toward Buck, Angelo, Bill, Tim. They know all these men well. They treat them all generously. They ask about details of wives and kids gleaned from past luncheons. They recall outings sponsored by their company, on which everyone had a good time. They are the source of new long-billed caps that everyone has taken, with chemical-company names, and they are a source of belt buckles. The salesmen complement one another. One is short and pockmarked, near retirement, brusque. The other is tall, has round rosy cheeks with a tiny brush of a mustache connecting them, and a mild, passive, amiable manner.

The salesmen deal out two other assets to their friends at table. One is respect. The routine, line-of-duty purchases of fertilizer, pesticide, herbicide that Buck will make are budgeted and approved. He does have the power to say which, and, providing his decisions are routine, the power to say when and how much of which goes on his plants. He can spend a lot of money as long as he risks none and doesn't spend it on surprises.

He enjoys the position, although the style of manly conduct out here calls for a show of fair play and a manner that betrays no comprehension of his own power. Respect is traded. It's bad form, at least until one is much higher in the bureaucracy than Buck is, to betray an awareness of one's own strength over pals. When Buck goes bowling he turns away before the ball strikes the pins, and doesn't smile when he scores a strike. As he puts it, he "gets enjoyment out of" hunting, bowling, playing with his kids, making fertilizer deals. Out of respect for himself and his companions, he just doesn't admit it. The fertilizer salesmen just happen to be selling fertilizer and giving away caps and belts to their good buddies.

The other thing that the salesmen have that Buck needs is the latest news. They know how "the substance," the chemical now under discussion, has worked over at the J. G. Boswell place, how "the material" (it's just come out but is already worth considering) did up at the Tenneco farm and over at the Heggeblade-Marguleas farm.

The short, older salesman and the tall, younger salesman help Buck to cover his rear. And that — even in the absence of the companionship, outings, respect, belts and caps — that alone would be enough to convene the meal. So they all pretend they are not working, are not competing, are not under pressure to do what they are doing. The rules make it hard for an outsider to follow the game. It's difficult to know who's trying, who cares, who takes care, who is trustworthy, who is friendly, when everyone pretends at all times to have perfect control of every situation and not to be bothered by competition. It seems a wearing world in which to live.

"Here I am," Buck complains to the assembled crowd, "harvesting day and night. That Etheril you guys sell me makes everything a rush — you fly it on and the tomatoes ripen with a bang and then go on by. I'm tired. I'm going twenty-four hours a day. I tried to play with my baby boy this morning — and I fell asleep with him climbing on me. I'm harvesting while he's home without me. I miss a whole month of his life. It ruins the enjoyment."

Lunch breaks up; new caps are gathered. In the parking lot, Buck and the salesmen drift off toward Buck's pickup. The young big one lists points on his fingers. The older, softer one nods reassuringly. Buck shrugs, palms up. They shake hands. Everyone heads back to work.

Buck and I drive over to the company's airstrip. It's miles up the highway, but it might as well be next door to Denny's — there's no change in the glare of the pale flora or flat terrain between the two locations. Yellow sand bands the roadsides. Miles of perfect crop rows flicker on the flat ground. Geysers of irrigation water, feathery lines of transparent plumes, dark wetted lanes of dampened soil all converge at the horizon. Everything quavers in the heat.

The airport is only a thin strip of roadway by the side of the real roadway. A windsock dangles from a steel pole. In the open shed next to it tall stacks of bagged pesticide lean against each other. Buck squints at his watch, then points a whole arm as if he's sighting along a rifle barrel. He points out a distant plane approaching. It is very small, still far off. But it coasts toward us, just over the crop rows, like a tractor with unusually high ground clearance. The noise when it lands sounds like a whole stock-car race. The plane is stubby and fat like a tractor, and its propeller seems far larger than a single-seater ought to have.

"It carries tons of material on every ride," Buck explains.

The pilot heaves himself out of the open cockpit. He is heavy-set, with a smudged face, genuine aviator goggles, and a quipping, fast-talking, chummy manner. He was struggling to make it in a competitive business a few years ago, Buck tells me, when he landed a fat contract to do Tejon's spray work for them. He is most obliging. He smiles, stands waiting.

"I'm going to bring you down a truckload of Monitor for the cotton by the tomato harvest," Buck says. Perhaps this is the result of the chemical salesmen's luncheon labor. Buck signs a work order, lists by number a couple of square miles of fields that will get sprayed.

The pilot disappears into his shed to "check his schedule." He returns, carrying a corrugated cardboard box. Leaves and dried grass hang out around the edge of the closed lid. Opening the top just enough to admit his hand, the pilot draws out a long spotted snake. He whoops and whirls it around while Buck looks on, alarmed.

"He's my new pet. Found him right over there," says the pilot, pointing.

"Treat him well," Buck says. "He's a gopher snake, a pest predator."

"Christ, I ought to kill him quick then," the pilot replied, "because he's going to cut into my business. Maybe I'll be a snake charmer when I go out of business. Sing the oy-vay song and he'll rise up."

Buck smiles, the pilot laughs out loud. He puts the snake back in the shed, lugs out sacks of poison and loads up his tanks again. As he works he continues chatting. He agrees to spray the cotton the next morning. His daughter knows where the fields are. She has maps. She "flags" for him — stands on the ground next to the fields he has to spray, waving a white peace flag so he doesn't deliver the wrong cloud to the wrong crop. The pilot climbs back aboard and takes off, banks sharply, circles, then flies away, waggling his wings.

"I'll tell you something the pilot told me that he saw last week — you'll probably make something of it," Buck says as we drive away. "He said while he was flying a load off across the place, he looked down, and one of the guys down there who was working up a field had driven the tractor around and disked his name into the field, in letters a hundred feet high. 'José,' it said."

Buck says he spends his days putting out fires. He isn't due at the tomato harvest until evening. He says the day is going normally — always trouble, always something different. People trouble, machine trouble. But he guesses he likes it well enough.

"When you're asshole deep in alligators, and all of the sudden you see things click and work right, that's a high. There's nothing like it. It ought to happen more here, but there's so much bull going on to ruin it. I guess I think my job here is like being in a weight-lifting room — you train up, learn for good jobs later. I don't guess I'll own a farm. But I sure would like to work where they'll let me go ahead and make them some money."

We arrive back at the field office where he has a desk. Constructed in the midst of TAP operations, the building is low, prefab, steel the color of the desert sand. A tall chain-link fence rings both the field offices and Angelo's maintenance shop. The long corridor is alive with the work of the day. The gentlemen in the luncheon circle rush past each other as if they've never met. Phones ring. Equipment salesmen, more chemical salesmen, seed salesmen, acid salesmen (some of the soil is alkaline and gets sprayed with sulfuric acid), miracle soil-additive salesmen, tool salesmen, tire salesmen, irrigation nozzle salesmen, all wander through, samples, belt buckles and giveaway hats in hand, making the place a sort of reverse bazaar; here the merchants wander and the customers stay seated. Coffee is still the coinage of graciousness. The managers are always interrupting crucial sales talks, holding up fingers, palms, arms to halt conversation while they take up new conversations over the busy telephones. The receivers all have shoulder brackets on them. It is a building full of people representing things as they are not — salesmen representing great products and reliable outfits that think of the customer first, and TAP's managers representing the best interests of the customer, although the customer in this case has interests that are difficult for team members to decode. As I leave, I hear Buck on the telephone with a purveyor of something or another, embodying corporate welfare. "We were real pleased with the last batch. Bring us . . ."

A couple of freeway exits nearer to Los Angeles, just south of the last field in the Valley, at the beginning of the "Grapevine," the ten-mile long chute of freeway that twists upward through the Tehachapi Mountains, Tejon has constructed yet another office building. This building is also low, but is made of stone and redwood and glass. An inner courtyard provides offices with manicured viewing. Johnny Riley works here — this is his station when he is "so busy with paperwork I don't get out as much as I'd like to into the field."

I see him through glass, working down through a stack of documents. At the time of this visit, he is a high-ranking field officer at TAP, soon to be promoted. The reigning number one, a gruff stocky bachelor, fiftyish and sporting work boots, is a seasoned corporate farm manager named Leigh Heath. He is about to leave; I catch up with him only rarely. I pass his office, which smells of cigar smoke, and see him on the telephone. He may be talking to a salesman, or perhaps to a contract harvester or a labor contractor. "Well, when *can* you get them to us?" is what I hear him roar into the mouthpiece.

Some of the interests of Tejon Ranch other than TAP also are supervised from this office. The commercial and land use division, which exploits the business potential of the ranch's five consecutive freeway exits, is overseen in part here. The ranch has constructed and leased to tenants eight service stations, three restaurants, a car repair shop, a post office, a farm market, an office building. This division of the company also carries much of the burden of the high land taxes on Tejon's thirty thousand acres of Los Angeles County that are being held for future development potential.

Early in the 1970s, Tejon planned an eight-thousand-acre development with lake, golf course, tennis club, stable, house lots and condominiums. The Sierra Club filed suit, according to Joe Fontaine, a local environmentalist active in the club. His objections were based on what he termed the development's "lack of acceptable plans for water and sewage supplies, lack of plans adequately considering fire, police, school costs, environmental effects, and the earthquake fault-lines passing through the area." Because of the neighborhood opposition, the suit, as well as inflating construction costs, an end of an earlier resort development "boomlet" in the area, and changed governmental regulations, the project was "postponed." Somewhat sheepishly, Joe Fontaine says, "The demand for that sort of thing peaked and passed. They are probably down there thanking us for the money we saved them by opposing this development." In conversation, some years later, a Tejon Ranch Company vice-president ironically praised the Sierra Club suit on just such grounds.

On my way to this building I had driven the length of Tejon Ranch's land and had passed hundreds of oil wells. Some are silent, some repeat dismal groaning sounds with every stroke. Each well is pumped by a long crossbar, the well shaft fastened to one end by a large shaft guide the shape of a grasshopper head. Counterweighted legs swivel from the crossbar's other end. Clustered in the distance, the pumps look like a crowd of locusts, salaaming. It's said to be a dollar a bow; oil has been the strength of the parent company, although its land base seems to be taking precedence now. The sheikhdom's oil trade is managed now by just one vice-president. He is also responsible for overseeing collection of royalties on the cement plant General Portland now operates at the site of the ranch's limestone deposits. He also frequents the Grapevine office building. He oversees the leases (to Superior, Exxon and Gulf among others) under which other oil companies mine the ground. As in the coal areas of the Alleghenies, it's common to find the land around Bakersfield deeded in two layers — conventional land users hold the top five hundred feet, and large subterranean landlords, usually the big energy companies, hold title to the riches that lie below. The Tejon Ranch vice-president, Oil and Minerals Division, was out as I passed his office. Behind his clean desk hung maps with small colored flags marking well sites.

Ray Peterson works near the end of the long hallway. At the time he was still a vice-president of Tejon Ranch Corporation. He was also vice-president and on the board of TAP's general partner, Tejon Agricultural Corporation (on the TAC board with Peterson were several investment bankers, the president of a major cattle and land company and the director of a University of California Agricultural Experiment Station). Ray Peterson's office is spare — a desk, a shelf of books, walls covered with charts. Peterson says he stands outside the line of command and advises. He's the company economist, a fortune-teller of future supply and demand.

He is tall, fiftyish, angular and stooped. He sports a dead-level flattop and has a military bearing. He may be the world's last surviving Social Darwinist. "With varying degrees of intelligence," he says, "there are varying ideas of one's best interest." He's among the cleverest men I have encountered in my journeys to this farm. He's defending the size and power of the operation, and seems amused by the workout.

"Management is misunderstood — underestimated in farming," he says. "Imagine that instead of the Tejon Agricultural Partners group, whose interests I help supervise as part of a skilled team, there were a hundred-thirty-odd family farmers — each with a hundred sixty acres — and each of them was farming independently. There would be lots of competent tractor drivers, good hard workers not too good at managing, and there they would be, managing. The way we have it arranged, people are out there doing the things they are skilled at." He smiles. I can't tell if he really means it. What if, I ask him, of those family farmers, there were not one but ten fine managers who could also do every other farming thing well, and *they* had the land — wouldn't that be a better arrangement because it would maximize the free market aspect of farming?

"It would seem that way, but their land use wouldn't be efficient. They would grow what they felt like growing on what land they had — and our soils differ very much. Crops would not be assigned to land they're best suited to. There's an optimum crop pattern. You never hit it. It's how close you come — a measure of success. I guess there's an optimal management allocation too. . . ." He grins.

"Our farm general manager has his job because of his superior organizational talent. I have the ability to help run the farm," he explains, "so I help to run it." He shrugs a shrug that imparts to this confused observer the clarity with which he sees obvious order in this chaotic world.

Having been instructed by Ray Peterson to keep my eye peeled for "superior organizational talent," I rather willfully keep my eye peeled for a farmer. I spend days combing TAP's twenty-one-thousand-acre premises for one.

"Are you a farmer?" I ask middle management. No one feels he is, except for Angelo Mazzei, whose house sits on twenty acres of farmland just inside the Bakersfield city limits, whose brother brokers vegetables, and who consequently plays after working hours with growing a field of sweet potatoes for market. "Yes, I'm a farmer," Angelo says, "but not at Tejon, not here."

"Here," middle managers feel they are middle managers, or shop managers, or field crops managers, or irrigation supervisors. I asked some of the people they supervise, workers who drive tractors, move pipe, plant, hoe and pick crops, touch plants, the same question.

"Are you a farmer?"

"Of course not."

"No, señor."

"I work with crops, but I don't decide what goes in, and I don't get the profits. I'm just a damned wage earner."

"I'm a people manager."

"I dunno. Never thought about it."

Weeks later, I drive away from the Tejon field offices, out of the hot desert valley, up the Grapevine right past the partnership's headquarters building, and continue southward across the mountains into Los Angeles, down Wilshire Boulevard. I park, then ride an elevator up a dozen floors of an aluminum and glass office building, pass through a commodious lobby. On the lobby wall hangs an artwork that seems to be an abstract map, done up of textured leather, sheet metal and upholstery tacks, of the entire three-hundred-thousand-or-so-acre dominion of Tejon Ranch Corporation. In a corner office possessing a fine view of Los Angeles still unveiled by smog, I shake hands with a man named Howard Leach, who is now gone from his job — at the time he was president of the Ranch Company and chairman of the board of Tejon Agricultural Corporation, general partner of TAP. The hand is large, the handshake strong and directive. I sit as urged, in a chrome and black leather easy chair and greet Leach, about whom Johnny Riley has said, "He is a marvelous director; he motivated this ranch — it was dormant until he came along." Leach wears an impeccable blue seersucker suit. He introduces me to a lawyer, employed by the ranch, with whom he's been conferring. They chat for a moment more, then the lawyer leaves. The telephone rings. Leach gestures just as avidly to an absent caller as he has to the corporation counsel.

"Where are you headquartered?" he asks as his pointing finger brings home the immediacy of the question. He listens, nods, and says, "Well, in that case I could fly down to Ventura and meet you at noon." Leach hangs up, makes another quick call and discusses an arrangement to helicopter off in the opposite direction the following morning. Finally he turns my way.

"Are you a farmer?" I ask.

"Hell yes, boy!" He looks surprised, injured that it isn't obvious. Hanging on the wall behind him is another picture, brown and dark and very representational — not like the map in the lobby at all — a little one-horse farmstead, wind- and water-lashed, in the midst of a thunderstorm.

"The basic problem of any farming enterprise is the cyclical nature of agricultural prices . . . ," Howard Leach explained during our interview. It's a truism, of course, so general it is not of use in determining why some farms succeed while others fail. But at the same time, the observation provides a background for understanding the goings-on at TAP. Some of TAP's trouble was indeed "historic" — they were doing what only hindsight can reveal to be the wrong thing in the wrong year. When Tejon started up, buying many millions of dollars' worth of supplies and services, America's agriculture was entering an unprecedented boom time. Farmers who had eked out livings from their ground for a lifetime made killings in 1973.

The problem of boom years is that farm "input" prices follow rising crop prices upward. It's called a ratchet effect. The year after a boom, equipment and supplies cost more. TAP had no boom-time crop, but it did need to purchase equipment. At the same time, boom prices encourage new entrants willing to compete for room in the attractive marketplace. The result is, of course, that once demand is satisfied at the high crop price level, prices start to fall. Costs of supplies sometimes don't. Thus agriculture's cycles, and, in general at least, TAP's ensuing difficulties.

The planners of TAP had projected that 43 percent of the partnership's ground would be put into vineyards and that they would eventually be one of the most lucrative parts of the operation. These plans were made, undoubtedly, with full knowledge that tax advantages and high prices for grapes were attracting many others to the business of developing vineyards. America, though, seemed to have just discovered wine, and demand for it rose as quickly as supply increased — at least for a time. TAP's prospectus quotes Kern County grape prices (drawn from government statistics) as averaging $116.65 per ton, although it warned that they might not stay there. They climbed to $151.00 in 1973, then fell, in 1974, to $95.00. The prospectus also made guarded reports — again based, no doubt in good faith, on government statistics — of how quickly the fledgling vines would mature enough to begin yielding payback tonnages. What happened next may well be attributable to "the cyclical nature of agricultural prices." Grape prices declined in successive years as new yields competed with old ones. And Tejon's vineyards were reaching maturity "about a year later than projected," according to a ranch vice-president, tapping into a steadily less rewarding price structure. The prices have recovered today, especially for white wine grapes, but they gave little cause for optimism in 1975. That year, Howard Leach wrote in TAP's annual report to the limited partners, "An improvement in wine grape prices is important to the Partnership since this crop has always been expected to be the major contributor to our revenues and the effect of inflationary increases in costs of land preparation, improvement and cultural care, the Partnership has made substantial borrowings. . . . The need to service these obligations underlines the importance to the Partnership of a successful wine grape marketing program." Prices continued to fall. Yields continued to arrive at projected points a year late. The need to "service obligations," namely, outstanding loans, as well as fees due the general manager and the parent company for management and land use, continued.

The table, from TAP's July 1972 Prospectus (footnotes omitted here) details the basic monies that flowed at once from TAP to Tejon Ranch and Tejon Agricultural Corporation, its subsidiary.

FARM MANAGER
Payments for farm assets $2,351,000
Reimbursement for farm preparation
    expenditures 649,000
Reimbursement for calendar year
    1972 planting preparation 1,500,000
Farm management fee 1,070,000

GENERAL PARTNER
Guaranteed payment on land contribution 1,150,000
                                                       $6,720,000

The first three items were one-time payments. The fourth item was to be annual, but reduced to about three quarters of a million dollars after 1972. The fifth item was to be annual. By placing its reimbursements and payment for assets at the beginning of TAP's life span, the parent company undoubtedly reduced its risk — monies they put into the project would be recovered rapidly. The payments did add materially to the "cash outflow" during the early days of TAP.

Another obligation that needed to be "serviced" — in this case, serviced come hell or *low* water — was the agreement that committed TAP to paying the high fixed costs of receiving water, charges that continued even when the drought hit California and there was too little water to go around. In the drought years, TAP had both to pay those charges and also pay, when enough canal water wasn't available, the costs of pumping underground water onto their thirsty crops — a double expense at a time when the "cyclical nature of farm prices" suggested thrift.

"Harlow Bulware," a foreman in the vines and trees division, puts some blame on freakish weather. "They were already in trouble. Then drought came. And then this wind and sandstorm came right through. Made gullies in the roads and fields two yards deep — it looked like an earthquake. It wiped out nut trees and a citrus grove — nearly a thousand acres, back just below where the sand came off the mountains at the edge of the valley — wiped it out so completely we just shut off the water and let the investment go. There were some staff houses back there. It smashed every window in them, then peppered the interior walls with rocks like a machine gun had shot it up. Some cars parked back there were buried. Others were sandblasted until the windows were etched and the bare metal shined. The next day it was sunny."

The cars still sit back there, eaten by rust, half buried on the edge of the six-foot gullies that haven't yet been repaired, next to the dead orange trees.

Several times, I came to Tejon after staying with farming friends in Iowa, and the contrasts in styles of operation seemed especially marked. The midwesterners farmed nearly perfectly. Their machines were always greased and shining, their crops as bountiful as weather allowed, their improvisations with old buildings and materials at hand reflected the elegant sorts of accomplishments farmers manifest when thrift guides inventiveness.

On the big corporate farm, everything was done as per order, and, quite naturally, reflected the weakness of applying military order to the growing of crops. The contingencies weren't covered. What to do next when what happened last didn't work, or hadn't come off as per schedule and as per budget, wasn't obvious. There did not always seem to be much staff interest spent keeping stray details in mind. Divers departments were to see to divers chores — the situation occasionally reminded me of big-city construction trades — carpenters standing idle while bricklayers lay bricks slowly. It seemed as if the corporation could buy workers' time, managers' time, planners' time, and even all of their skills, but that to buy a human's sense of responsibility is beyond the reach of big business. Like a lover's love, a prisoner's acquiescence, a parishioner's faith, or a friend's generosity, a worker's feeling of responsibility — of compliant but sensible allegiance to the completion of the chore at hand — simply never comes onto the open market.

The impulse to do a careful job, to treat work as craft, arises from some private preserve of wholeheartedness, of willingness to attach self to action. Many workers at all levels did work wholeheartedly; others withheld some commitment — frequently with what appeared to this observer to be understandable cause. The large body of modern management theory having to do with how to make workers feel respected has arisen from efforts to invade this preserve. Workers, at least workers on big corporate farms, can be counted on to know what makes sense. The instinct to conserve creature-energy that has cats napping when they aren't hunting also protects workers from giving their all where it seems chancy, humiliating or pointless.

"First of all," the former farm manager, Leigh Heath, once said, "we handle any decision at the 'first level of competence,' the first person competent to make a decision gets to make it. Second of all, we're a team. If a person has the right information and training, he shouldn't have to see three people to do his job. Committee decisions are used, but only in cases that cross expertise."

Johnny Riley had said, "A person is successful if he surrounds himself with good people."

John Wood, the current farm manager, says, "At whatever level, I view the people here as presidents of their own little companies. If you have good people, that's the best way to get the most out of them, and we have good people."

The seeming contradiction between claims of managerial autonomy within a teamlike structure and the sharp feelings of betrayal and vulnerability managers may feel about their bosses reflects the stress that propels management strategy along. It reflects a structure in which every person above the ground floor of the hierarchy is pressed and made responsible for all that happens in his purview. The stress is transmitted downward and upward. It generates, inevitably, a style of decision making that characterizes modern corporate farming. The decisions come out conservative, sound, uninspired, and thorough. Nothing is accomplished with flash. Nothing is done cheaply. Failures are all complex and have to do with systems and processes that are barely understood by those involved. The company's mission is slowly and continually accomplished, rather well but not very well.

Early on one visit, made at an especially demoralized time when one legion of managers was about to depart and a seemingly more successful crew was about to take over, I drove far back through Tejon Agricultural Partners' land, far from the field offices, together with a new foreman who had been sent out by his supervisor to check the progress of a labor crew. A tractor driver saw the company pickup approaching, pulled to the side of the road, and flagged us down. He wanted advice. He'd been sent out to disc-harrow a field, row crops division, and the harrow needed a little adjustment. This was clearly a problem for the crops division to handle. The foreman was from grapes. The foreman got on his radio and called in the field crops supervisor, who said he'd check after lunch. As we drove away, the foreman asked me, "Did you see the oil dripping down the side of the tractor motor? Looks like a leaky gasket. Probably down on oil, too."

"Why didn't you report it to the maintenance shop?"

"Why call it in? Nothing'll happen anyhow. You know what our situation is at the moment? Four of the sixteen tractors our division can use are in operating condition — some of the down ones have been down for a month or more. We're on overtime and night shifts, just so we can get our work done with the tractors that still work. That's expensive, too."

"Another time," a former employee recalled, "there was this big tractor, a Deere 7520 four-wheel drive. It ran low on water because the guy driving it didn't notice. It heated, blew its rings. Maintenance said they'd rebuild it. They took it apart, then the guy doing it left. The parts sat around the shop, getting messed up. The shop called the tractor dealer and said send over a mechanic, but don't take it in, just pick up what's here, and put it back together — there's only so much money in the budget. Well, they sent over a guy and he did just what he was told. A few weeks later the tractor was down again, because too many people worried about little parts of their job, and their budget, and no one had taken hold of the situation and done things right. It had to be rebuilt again. At the dealer's shop this time. Cost four grand to do."

On still another ride, this one with Buck Klein during his days as row crops supervisor, on the way to check on the garlic harvest we drove past a sprinkler line on a section of cotton field already sodden. Cotton, given its way with water, will grow ten feet tall and never set seed — and the seedpod is the cotton part of cotton. We passed a flooded field. There must have been fifteen or twenty acres being overirrigated — at three five-hundred-pound bales to the acre, say sixty bales, just over fifty cents a pound, eighteen thousand dollars' worth of cotton soaking and in jeopardy. The foreman reported the situation over his two-way radio. Someone in the irrigation department responded, "Thanks, we'll get to it tomorrow. It's late in the day now." Buck shook his head in amazement. It is hard to imagine a family farmer in Iowa — or California, for that matter — letting this happen.

After checking the garlic harvest, we drove on to a field of cannery tomatoes slated to be abandoned, and not harvested. There were fifty-five acres of them, at a field still called Guidara — from the name of the farmer who had leased it and made his living there before Tejon Agricultural Partners called it back in. "Trouble here," Buck said, "is that the tomatoes are so small. There are plenty of them, but they're like goat turds. It wouldn't be worth the money to bring them in. It's a writeoff. Tomato land is budgeted at about eight hundred dollars an acre expenses, and whatever eight hundred times fifty-five acres is is what we lose here. The thing is, this is not tomato land. I knew it before we started, and told them. It's too sandy. But that's how they wanted it."

On another tour through the fields. Buck took me back to the site of a carrot harvest. In the world of large-scale corporate agriculture in California, it turns out that all harvests look about the same. The shape of the machine varies; the sparseness of human labor varies slightly. The form is the same. Eighteen-wheelers gather in a staging area next to the flatness currently being worked. Oddly constructed harvest equipment does something special. In the case of carrots, the awkwardness of the machine includes protuberant wings, and a little platform on the back where someone sits, aiming some mechanized rooting about. The result of these manipulations (not to mention those of ag engineers, steel fabricators, extension researchers, bankers, and the like) is that two rows of harvested carrots, dug from a forty-inch-wide carrot bed that proceeds toward the horizon in parallel with thousands of others, wind up in a big hopper. The hopper winds up in the assembly yard, and thence behind a truck that lugs it off somewhere to a processing plant where the carrots are cooled, washed, topped, and packaged, all automatically. "The plant owners make more from our carrots than we do. These here are jointed — I think with Yurosek. Market price now is about four dollars a forty-eight-pound package, and our break-even is three dollars, at sixteen tons to the acre. We aim for seven hundred fifty thousand carrots to the acre — seventeen carrots to the foot. It's not a high-profit crop, usually. This year, there will be a huge loss. Why?"

Buck points silently across a sandy road to another field of carrots, slightly bigger of top, and looking extremely healthy. A crawler, not a harvesting machine, makes its way through the field. Behind the crawler is hitched an enormous set of harrows. As it moves along, I can see pieces of chopped carrot, carrot tops, carrot tips, churning in its dirty wake. The ground settles down clean where the machine has passed. I ask the obvious question. What was wrong with the carrots — why couldn't they be harvested?

"There are enough carrots on the world right now without these," Buck said. "Price isn't so hot, and the warehouses were full when these got to be the right size. We were held off harvesting. Someone let time go by and suddenly they were too big. More than eighty acres of them, which comes to sixty million carrots or so. They couldn't fit into those plastic carrot sacks they sell carrots in unless they were cut, and that would have cost the processor a bundle. They offered us a hundred and twenty-five dollars an acre for the carrots — and it would have run us two hundred dollars just to have them contract-harvested. So this is the cheapest alternative. It seems a shame, though. All that food."

Days later, when I asked Ray Peterson, the economist, about

the failed carrot harvest, he just shrugged. "Vegetables are speculative crops to grow. If ever you want a proof of the fact that a free market still exists in farming, you just saw one."

However, the problem did not seem to be just one of low prices — which is the sort of difficulty producers do encounter in volatile free markets. Rather, it seemed to be one of coordination, of timing, of what may have been unclearly exercised authority for picking a harvest date, of an arthritic marketing structure beyond the corporate level.

Another problem altogether, or a series of other problems, seems to have caused a defect in a sizable portion of the thousands of acres of almond plantings. Some persons in charge either did not know how almond trees ought to be pruned, or were not attentive enough to pruning operations. According to experienced nut farmers, almond trees need pruning while young if they are to yield well, and if they are to grow strong limbs that will support the weight of the mature nuts. And the way to prune almonds is to create a series of angled forks — trunk into several branches, each branch into several more — all separating at less than right-angle intersections. One wants one's trees forked in acute angles, like a hand held with fingers splayed, not boxy like a candelabrum or a family tree. Right-angle joints are weak and can crack under harvest loads and in winter storms. At Tejon, some of the trees were said by some of those involved to have been pruned too boxy and squared. And many of the trees were said to have been pruned far too severely.

Karl Fanucchi, the young farmer who managed Tejon's row crops early in his career, before taking over management of River West, a large and well-run private farm north of Bakersfield, recalls his frustration with the almond pruning work. "I'm a nut man. That's what I like best — you can see it develop, and think it's your work. I had worked in many nut orchards before even though I was in field crops at Tejon. I saw what they were doing to the almonds, and I went to the men who were overseeing it then, and I told them. But the operation was in the hands of others at that point, so there was nothing I could do."

Another incident, that reportedly took place in the almond orchard, was described to me by middle-management people who were around at the time. It seems that a manager with some authority in the division of TAP that supervised the nut trees decided that hundreds of acres of nut orchard were weedy and wanted spraying, at a total cost of about twelve dollars an acre, with an herbicide. After this was done, someone who perhaps misunderstood orders or usual practices caused the orchards to be cross-cultivated so deeply that soil was raised to the surface from below the zone penetrated by the chemicals. As a result, weed-control measures had to be started again, at great cost to the partnership.

The long-range effects of the pruning errors are still felt at the ranch. Current farm general manager John Wood said that as a result of the faulty pruning, the trees have been set back in their development by at least a year — and at a time when the company is hurting badly for cash. The bearing branches in the pruned almond orchards have been tied back now — bands of tape encircle each tree. The operation cost "about a quarter a tree" and will have to be repeated at intervals during the trees' yielding period. Pickers sometimes cut the bands, because they hinder rapid harvesting. Wood said that the trees were yielding 500 pounds in 1978, but would have been up to 1,000 if the pruning had been properly executed. If the pruning setbacks were true for all the almond acreage, at nearly $1 a pound, $500 more an acre on 3,915 acres of almonds would be a considerable aid to a meager cash flow.

The spectacle of small "errors" of farming practice seemed commonplace at TAP. Visits to Tejon — during the era of their near demise — left me feeling like a village priest on the prowl. "Oh my!" is what I thought.

"What's that geyser of water doing spraying up over there?"

"Too light pipe was used for this part of the irrigation installation — an air vent blew off. They do. It was a mistake."

"Why isn't that section — that mile square of land — planted to anything?"

"It takes time and planning and capital — we have plans."

"How come that big field of cotton is so foul with Bermuda grass?"

"It was a wrong choice of crop. With that weed, it should have been summer fallowed or put into grain."

"Why are there holes in the cotton field there?"

"We've got around a thousand acres of cotton this year — but only an eighty-five percent stand on much of it. The people watching it weren't. It's not uniformly planted. It's not ripening uniformly. The planters were too deep."

"Those stakes and wires in grapes — they must have cost a fortune."

"The stakes are hardwood from Malaysia. When we were first putting in this vineyard, some of us were supposed to look around and make calls — find the best deal, because we were buying quantity. I remember one deal, we found if we moved the same week that we could save sixty to seventy thousand dollars. The 'yes' came down a month later. By that time, some real farmer somewhere had lucked out on the deal. That's just one decision. There were so many decisions that came back too slow."

Karl Fanucchi, the former row crops supervisor who has moved on to River West farm, is a popular man. He's big, quick-witted, a quipper. His broad face frequently reddens with laughter. He nudges, waits for response, gestures broadly with both his hands. Farmers around the valley call him when they are puzzled. He's the favorite of the most knowledgeable in the good-old-boy network. Although only in his mid-thirties, he has a well-stocked store of specialized agricultural knowledge, and of business knowledge, and of the skills needed to gain the allegiance of hired managers. "It was a loss to Tejon," a long-time U.S. Extension agent commented to me about Karl, "that they didn't keep him. Any farm that gives him a free hand doesn't have to worry about making ends meet. He's good." Karl Fanucchi's memories of his years at work for Tejon Agricultural Partners are memories of frustration, although he laughs as he tells about them.

"You always had to be accounting to someone for why you did something four months back. One time, end of harvest, I'd joined three separate cotton co-ops to get my ginning done. All gave good service, and it helped our neighborhood relations. Got us the best price too. In the middle of ginning, someone calls from Los Angeles, calls me down there. I go in, and he asks, 'Why are we using three co-ops?'

" 'Because it's the best decision,' I say.

" 'Do you have the authority to make it?' they ask.

" 'Yep.'

" 'We don't know if you should have made that choice,' they say."

As the former supervisor concludes the story, "I shrugged and I left the meeting. And I kept on making decisions. Here's another.

"It was around the time of the incident with the cotton gin co-ops. I spent half a million on fertilizer, and was called down for it. One boss was aghast. He was eating the ash from his cee-gar. It saved them a quarter of a million dollars to buy it when they did, and I got called down for doing them a favor. That's how it was down there. It's progressive farming, you know? That place didn't compete on efficiency. I saw wastes that would make you sad. I could document so much sheer waste. Fancy farming by committee!

"There's stuff planted down at Tejon that I put in. I still go and look at it. I feel like there's a part of me down there. At

River West, I'm busier than a cat covering crap. It's nonstop, but whatever needs doing you can get done. Down there, I had to leave."

"Oh my."

Some of the troubles during Tejon Agricultural Partners' development years were difficulties staying organized and farming efficiently. Others involved treating employees, such as Karl Fanucchi, tactfully enough to keep intact the will to work hard. Buck Klein recalls one incident in which what was felt to be due was not forthcoming. "I had people — workers — bugging me for their raises. They'd been approved in July. In September, after much hard feeling, they finally came through."

One middle manager, who we will call Jimmy O'Brien, impressively trained and especially active in organizing and supervising TAP's vineyards, earlier on seemed to feel chronically unrecognized. "We won't ever be powerful here," he said. "Upper management always has to have the glory, but they always want somebody to blame. I do the best I can for the company. I have no idea if they appreciate it or not. Everybody likes recognition, though. I write letters of appreciation to my staff when they work extra hard. I learned it at the Dale Carnegie course. But no one at TAP ever sent me a letter of appreciation. My satisfaction lies in my work, in the beautiful grapevines I am growing. My budget is for millions. I'm seldom patted on the back for saving them money. More graciousness would make working here pleasanter. But they don't think of it. Here, a man is a number."

It's a familiar complaint, and has been for most of this century — that man is a number, that respect isn't paid where due, that the work goes on anyway, because there is nothing else to do, that it's frustrating to keep on working in those circumstances. The attitude is generated by the rules of the game. One's performance in fulfilling a corporate job description is always under scrutiny, and there is always a judge, whose performance is likewise scrutinized. The relentless application of pressure and the consequent feeling of vulnerability replicate themselves again and again.

I am walking through the vineyards with Jimmy. The vines he has overseen, Thompson, Muscat, Pinot Noir, Ruby Cabernet, Chenin Blanc, Barbera, Petite Sirah, are really quite beautiful. He does things right. He finds time to be in the field. A small section of vineyard across the road that has not been under his care until this year still looks spotty, stunted, and ill-kept. Jimmy's dominion is uniform, his vines strong, precisely trained, the leaves a deep rich green. The plants quiver with life. They seem to have more energy than is necessary merely to survive, like teenagers at a dance. They are beginning to bear, but must be carried for four, five years from planting before they pay their own way.

"We spent eleven hundred dollars an acre for irrigation. Training vines for machine harvesting is more precise than training for hand-picking. The first year is just potted seedlings — we built the big greenhouse just to grow seedlings so that we could be sure they were virus-free. The second year, in December, after frost, we cut them all back to two buds — one extra, in case. The second spring, we take the shoot up to the stake, train it straight, tape it twice. It's got to go straight, not slack like a rope in the wind, because if it's slack it'll tangle up with the mechanical picker. That's why we lined the long rows up with laser beams too. Training is the art of making satisfactory permanent laterals. The first year it costs us a dollar and a half a foot. After a few years, only eighty cents."

We walk through the last of the three-year-old vines and come out into a sandy, open field. A crew of laborers is at work, setting up stakes for a new section of vineyard. A flatbed truck drives between the marked rows loaded with stakes. Every seven feet, men on the back toss another stake down onto the ground.

"It takes a lot of labor to plant these stakes into the ground, get them the same height, staple the wires on. We have a special post-driver truck — cost twenty thousand dollars, just to drive in the big end posts — they have to take the tension of the long wires. Drove forty thousand end posts with that truck, did some custom work too. It budgeted to a dollar and a half an end post. Paid for itself right there, and it's still good. The wire is drawn down the rows in spools. Twelve dollars' worth an acre.

"After the posts go in, the wire goes up. I put together — developed — another truck, has long flexible tubes coming out of a sort of trellis on the back of it, with pneumatic staplers on the end of each. That way, we just drive along the new rows, and workers walking behind can staple the wire up fast, two rows at a time."

We stop to watch a crew setting the stakes. Everyone is fully engaged. No one looks up at us. A small boy trudges through the sand after the truck, standing the tall stakes up in the holes someone has poked into the ground in all the tens of thousands of right spots.

"We budget sixty-five cents a stake. Eighteen for the cost of the stake. Two cents for scattering. Some for labor, some for management. The rest for installing. Each worker averages six, seven hundred stakes a day. They get six-and-a-half cents a stake, the labor contractor gets one-and-a-half. Piecework gives us speed, but lowers quality because it makes them rush. So each worker numbers the rows he does. Each is accountable to me that way for the quality of his work."

Perhaps the diplomat who sends letters of appreciation, who knows what it feels like to be treated as a number, who has ten square miles of creeping, tendrilly vines under perfect discipline, forgets for a moment. He has his job to do. A furrow creases his brow, amplified by the harshness of the sunlight. He strides to the end of a row, checks a number; he is at work. He waves his arm, beckoning the job foreman to him, and points at two rows of bare stakes. The stakes are unevenly planted. Some stand taller than others. He steps forward, throws a few of the worst cases of disorder down on the ground again. The dispersed crew now glances our way, but still keeps working.

Behind us, several rows away from Jimmy's conversation with the foreman, the stake truck has halted. The men aboard must not yet be aware of the situation. Two of them roughhouse on the empty truckbed. They laugh. They shove each other. Taking in hand two leftover stakes, they duel. They are pirates off the Barbary Coast, fighting for honor. They are not poor *braceros* who have been working hard on a dull hot afternoon, in line for a dressing down.

Jimmy hears the muffled clank of wood hitting wood, hears the laughter. He glances back, absorbs what is happening, but continues his conversation. The uneven stakes must be replaced properly. The company must not be charged for correcting this error. The two men smile. Then Jimmy turns and walks back toward the truck. Someone calls something out in Spanish. The pirates stop, look about, then stand waiting, grinning like guilty schoolboys. Jimmy's message is brief.

"No games," he says.

"Yessir," says the taller of the offenders. The two remain standing there with heads hung, as Jimmy turns and walks away.

"That's a waste of money," he says to me. "You respect them, they'll respect you. But worktime is not playtime."

All over the ranch, one can see crews of workers doing repetitive sorts of things, always stretched out in rough lines across the flat fields, always accompanied by foremen, subforemen, crew bosses. A farmer just to the north of TAP's vineyards employs prisoners on day-leave to work his vines. Perry says their work is so poor there's no point in using them.

Like most of the larger California growers — but not all of them — Tejon's management is extremely conscious of the company's potential vulnerability to organized labor. In a frank moment, former Tejon Ranch vice-president Ray Peterson allowed that in planting the crop array of Tejon Agricultural Partners, Inc., they had attempted to structure things so the crops would be relatively "boycottproof."

Stiffened into inflexibility by their size, big corporations plan their moves with lead time that allows their awkward mechanisms of communication to operate routinely. They provide to organized labor appealing targets of considerable magnitude. They are vulnerable to the leverage of striking fieldworkers in a way smaller growers never were.

Ironically, the high level of mechanization large growers seek, not only in pursuit of efficiency, but in order to bypass labor problems, in many cases only renders them more sensitive than ever before to strikes. Machinery eliminates workers, not work. It leaves each worker with a larger scope of accomplishment than before. If a single hand-tomato worker could pick — or refuse to pick — say, half a ton an eight-hour day, one tomato harvester driver can refuse to pick one hundred sixty tons in eight hours.

With fewer jobs, of course, the job market becomes a buyer's market. Tejon has managed, by means of caution, diplomacy and persuasion, to ward off attempts to unionize its workers. There is always a hungry worker willing to take what he can get to do the work — such is the decimation of the job market wrought on California agriculture by the age of high technology. The extreme vulnerability of large growers to organized labor helps explain the passion that Tejon's middle and upper management bring to their discussions of it.

I am sitting in Ray Peterson's large white air-conditioned car. He's parked by the edge of a field of lettuce. A dozen Mexican men and women stretch across the field in a long wavy row. They scrape weeds up with long-handled hoes. It's 101° out. The crew members wear kerchiefs, hats, white clothing. Their faces are red. They are too far away for us to see their features, but frequently one or another will straggle back, straighten and brush forehead with shirt-sleeves. It's obvious that they are hot.

"Our workers most probably would like something they don't have. That's a fundamental, a rule of economics," Peterson says.

At Tejon, most of the field labor has been performed by crews hired by a labor contractor. "We renew Joe Lopez's contract to supply us with labor because he's solved our labor problems,"

Johnny Riley has said about Tejon's labor contractor. Buck Klein has commented, "He gets us good crews."

Joe Lopez is a charming man, short, impeccably unmussed in a yellow short-sleeved shirt and khaki pants. He sports a thin, waxed handlebar mustache. He greets Ray, smiling, perhaps ingratiating. He gets 20 percent added on to what his crew grosses. "My profit is about three percent after expenses — insurance, accounting, counts on piecework, transportation. The contractor has closer touch with the people than the union hall. My workers prefer me. They see the union as an outside group. Labor contractors have an unfairly bad name. One rotten apple in a barrel doesn't mean all will go bad." He seems to be referring to Juan Corona, who has recently been in the news again.

Back in Ray Peterson's car, cooling off, we watch as the line of hoers advances. Peterson says, "Growers and farmers go back a long way in the freedom to own property, to compete in the labor marketplace, to succeed or fail on the basis of others' efficiency and productivity. In our case, a labor contractor serves as an intermediary between two opposed groups. If it becomes a monopoly, a union like the United Farm Workers interferes with free competition between parties involved. The contractor serves two masters. Right now there's a large number of workers with only agricultural skills. It's a pure supply-and-demand situation. High prices for the least pleasant work develop." The field crews, whose work must be less pleasant than Ray Peterson's, are paid at about the minimum wage rate.

Back at Tejon field offices, discussions of Tejon's "labor problems" ensue. The white management of several echelons have a go at revealing to the press the truth about farm labor.

"The UFW? You know what it is?" asks one vineyard manager, leaning forward confidentially. "I think it may be a communist front. That's what."

Johnny Riley, telling about his previous job, says, "I was growing a cantaloupe crop and there was union trouble. They'd stand on the road with bullhorns and state, 'If you don't strike, when you go home your daughter may be hurt. Your house may be burned.'"

Leigh Heath cleans up after Johnny Riley. "I'm not saying if Chavez was standing there they'd say that. I heard 'em say that while you're in the field your daughter may be raped. To that extent, the workers don't have free options. It isn't fair. I object to the union's misuse of force, intimidation, not giving people free choice."

The current TAP farm manager, John Wood, ignoring the effect that the threat of unionization has had on liberalizing the terms offered workers by employers holding unionization at bay, says, "We engage in a continuing fight to maintain our nonunion status. I feel the employee is better off out of the union. If he joins, he has two organizations to fight instead of one. Unions help out weak people."

More conventional wisdom usually has it that unions, especially new unions struggling to organize, help out strong people in a weak labor marketplace. There's some irony in the parallel rise of large-scale agriculture and the movement of farm workers to unionize. When most California vegetable farmers were rather poor, when they worked side by side with their hired employees — as many smaller California farmers still do — positions of workers and owners were more equal and disputes more personal. The separation of management from the physical acts of farming increased the political power of owners over workers, but this change also inspired unionization.

Many of the medium-sized growers, such as Lionel Steinberg, a farmer who harvests a thousand acres of premium table grapes a few hundred miles south of Tejon in the Cochella Valley, have welcomed the United Farm Workers and work successfully with them. Steinberg hires three quarters of a million hours of paid union labor each year. He has told me he has to bargain hard with them, and wishes for the "good old days" of cheap labor, but that the feeling of alliance and respect the organization has given the workers on the farm actually increases productivity. The large farmers in California all know of Steinberg, because he was the first grower to sign with Chavez. They do not approve. They feel their companies have too much at stake and cannot afford to lose control of their labor costs. They do not like the unions.

The issue is control and it's an age-old struggle. It has to do with pride and power and the profits of labor; it's always been nasty. Those in power, of course, want to keep what they have. In their comments on labor, members of the world of white farm managers frequently show an unworldly and naive disengagement, perhaps born of this fearful desire to hold *their* ground. Egalitarian attitudes toward competition with peers give way, during discussions of farm labor, to attitudes more bitter and ridiculing.

"Hey, you guess! What kind of crews harvest the watermelons around here?" one farm official asked me one afternoon while we toured the fields. He answers himself. "I'll tell you. Colored crews always harvest the melons. You guessed it. I don't know how, but they can always tell the ripe ones from the ones that need a few more days. Every time. One guy picks, tosses to a guy by the truck, and he tosses it in to another guy. They go fast. When they take a break, they always have a set of dice with 'em."

A TAP foreman philosophized once, as we strolled along the banks of canal number 850: "I envy the poor hoers. Eight hours and they can go home. Nice little house, your friends all around you. Why does a guy want a short life for another couple of hundred bucks a month? They get their breaks morning and afternoon. Nobady messes with *their* lunch hour!"

A shop supervisor praised a tractor for the simplicity of its controls. "Mexicanproof, this one," is what he said.

With a Tejon foreman, I once drove past an abandoned tractor-trailer body parked back between sand dunes far from the road. Two lawn chairs were set out in front of it. "Guadalajara Hilton," he said.

In a speech before the third annual meeting of the limited partners of TAP, Leigh Heath, then farm general manager, said, by way of praising the interracial harmony of the firm, "The University of New Mexico has asked TAP to train a carefully selected college-educated Navajo. . . ."

On stage the philosophy is more aware. "The way to avoid labor unions is to give the workers more respect than they ever get out of their union," Johnny Riley commented. "Also, to mechanize."

Ascribing to labor the motivation that Ray Peterson calls "fundamental, a rule of economics," seems correct. It's just that the "something they don't have that they would like to have" is something far more than money. What attracts workers to a rising union isn't merely the promise that militancy will lead to better wages. The allure must be that militancy may lead to shifts in actual control, shifts that would, in the end, identify laborers' services as unique, that would restore self-respect to lives now spent waiting to be eliminated from the service of unseen owners.

In the third quarter of the year 1976, Tejon Ranch Company discontinued payment to itself of management fees, then amounting to $35 an acre, or $708,737 per year. Its subsidiary, Tejon Agricultural Corporation, the general manager of TAP, decided to forgo its annual land use fee of $1,150,000. The cash crunch that showed up a year later was evidently on. In the 1977 annual report of the Tejon Ranch Company was further news of the difficulties afoot at TAP. "Approximately 10,000 acres of . . . land has been identified for possible sale to provide funds to sustain the operations of the partnership (2,700 acres sold during 1977)." The report also told of "material uncertainties as to the Partnership's continuation as a going concern."

Then, in February, 1978, Jack Morgan, who was still president of TAP, signed a formal letter to the John Hancock Life Insurance Company of Boston. The letter announced "a serious cash shortage" in the partners' operation — so serious that it would soon cause TAP to cease cultivating, fertilizing, watering and "otherwise caring for permanent plantings." In other words, TAP was no longer in a position to pay out the money needed to make the desert bloom, and Hancock officials were alerted that the collateral securing their $27 million loan stood in jeopardy.

Hancock decided to save the day. There were transcontinental visits of high-ranking executives — Rick Rickett, vice-president of Hancock's $1.5 billion ag loans division, came out, with entourage, to take stock. Certain persons resigned; others were asked to do so. There were executive talent hunts, replacements of key persons, negotiations between the parent company and Hancock, bargaining, revising. Ultimately, a draft was filed in the Kern County hall of records of a new "Loan Agreement between TAP and John Hancock Mutual Life."

Hancock would supply operating capital. The land identified for sale would be placed in escrow, some of it leased, some sold off as buyers were found. That would improve TAP's "debt position." House would be cleaned. The partnership would concentrate on its permanent crops and simplify its other operations. Leases of equipment would be reexamined, and restructured where possible.

A new corporate being, called Tejon Farming Corporation, would oversee the management of the farm. And a chain of agricultural consultants, paid for by TAP but working for John Hancock, would oversee the financial dealings of the farm far more closely than Hancock had overseen things previously.

In an office high in the Hancock Tower in Boston, late in 1978, Claredon ("Rick") Rickett explained the nature of Hancock's new connection to TAP. "We're running TAP *with* them — we're providing working capital. Through 1978, we've loaned in eight or ten million. Paul Teppen — a consultant we hire — he's there to help in any way he can, to see that our money's being well accounted for, that the various good practices are made. We are not managers — we supply money, and to protect it and keep it functioning, we judge managers."

Johnny Riley works elsewhere now. So do Jack Morgan and Leigh Heath and Howard Leach. Buck Klein has joined Karl Fanucchi at River West, and says, "I'm getting a chance to make some growers some money here." Angelo Mazzei now sells tractors for South Kern Machinery Company in Bakersfield, one of the largest John Deere dealerships in America. Ray Peterson also works in Bakersfield, selling real estate.

John Wood, the new farm general manager of TAP, takes a long view of the survival of such large farming ventures as the one that currently employs him. "With this type of farming, the original designer seldom survives. The first guy goes, because the financial burden is so great to put something like this into production. It takes years during which there is very limited return on investment and high continuing costs. The second guy may pick it up at a bargain price, but it still has management problems to work out. If he can't make it (and we're trying and I think we will), a third guy picks it up as the crops finally mature, and the cash flow is finally better. Then it works." Indeed, the "new regime" seems to have brought TAP through the difficult period and shaped it into a far stronger business.

Tim Heinrichs, agricultural reporter for the Bakersfield *Californian*, commented recently on TAP's difficulties during the reign of what John Wood calls "the first guy": "They planted the right thing at the wrong time — there were all those intensive capitalization costs associated with starting up permanent crops with modern irrigation and methods, and there were just too many other investors elsewhere doing it too. The book price didn't hold up. But I don't cry too hard for those investors who go in for this syndication business. There's a little greed there, a little greed involved."

As for those investors, their luck with their "units" has varied with their differing tax situations. An official at Tejon Ranch says that as of early 1979, investors had accumulated tax credits sufficient to recover their investments plus about 25 percent. And one limited partner, interviewed by telephone, said "I'd like to forget about that investment, but I'm close to even." A midwestern doctor who had bought heavily at the time of the initial offering says he has recovered four fifths or so of his investment in tax credits and doesn't expect to do better than that. A New England developer, on the other hand, who seemed extremely forthright, more than the others, had a different story to tell.

"I'm not optimistic that it's going to be a good investment per se," he commented. "But it's done very well as far as tax benefits are concerned. Unless the thing falls apart in a way that recaptures tax benefits — which could happen but doesn't seem likely — it will work out. I've made a lot in deductions, investment credits, water improvements — if you do it right you are filing a multipage tax form. I'd guess each thousand dollars invested has yielded three or four thousand in deductions. In its simplest form, this means that if a person owes a thousand dollars of taxes, by giving that thousand to Tejon for a year, you get to keep it yourself for the next three years." What it also means is that the taxpayers of America paid for a healthy part of the activity at TAP. When TAP eventually pays back its creditors, the IRS will consider partners' tax credits that came from developing the farm with loan money as yet untaxed as income. At that point the government may "recapture" some tax money previously retained by the limited partners.

Agricultural economist Theodore Shultz of the University of Chicago, the Nobel laureate who is the reigning dean of the discipline, in an interview with me once called giant farms such as Tejon Ranch "historically stabilized anomalies." He meant that there are some big farms — mostly dry-land cattle-grazing consortiums that struck oil — that have remained on the scene for enough time to qualify as "ongoing," but that in general, farms have not been driven by irresistible economic imperatives to combine into giant units of production. Shultz is right about farms the size of Tejon Ranch — three hundred thousand acres. But he might, in the light of very recent trends, have another judgment altogether about farms the size of TAP. In agricultural areas of the Corn Belt, the Wheat Belt, the Northwest, the Near South, the Southwest, the Far West, a noticeable minority of farms has grown into the ten-to-thirty-thousand-acre range. Since the advent of John Wood, the new general manager of TAP, the farm's operations have been greatly improved. It clearly is possible to run a fairly large farm efficiently enough to make money at it.

The question of what size of farm is most efficient seems to have troubled ag economists for all of this century. G. F. Warren wrote a report of a survey of Tompkins County, New York, agriculture in 1909 that included a discussion of farm size. He found simply that profits increased with size, because he studied only relatively small farms. In Black, Clawson, Sayre and Wilson's 1948 *Farm Management* the authors hint that there are upper limits to efficient farm size. "The best setup in many cases consists of a farmer as manager supervising his own work as farm laborer and very little more." Professor Earle Heady of Iowa State, in his classic *Economics of Agricultural Production and Resource Use*, writes: "It is doubtful that cost economies are great enough in most segments of American agriculture to endanger the units typically operated by farm families. Where these possibilities do exist . . . society may choose between larger farms as a means of attaining economies in food production and smaller farms as a means of attaining sociological objectives and political stability."

Heady unfortunately gives no hint of how "society" might go about doing anything as egalitarian as choosing between thrift and good politics, and the likelihood is that in the near future, such justice will happen only blindly. Our agricultural future is the vector of many wild and a few controllable forces in the world of economics, foreign policy, population, and technology. Americans do not seem eager to legislate farm size, in spite of the obvious relationship between how big local growers are and the nature of the resulting community milieu. We suffer severe inhibitions against interfering with the growth of entrepreneurs' equity, even in the name of greater social utility, and that attitude is not likely to alter soon.

Economists do generally agree that in farming, the "economies of scale" are in place at a rather low level — that for almost any crop, a farm with ten, or fifty, or five hundred employees will not have lower costs per melon, or per bushel of wheat, or per gallon of milk, than a farm with only one or two laborers. In fact, they agree that for almost any agricultural enterprise, the lowest "per-unit cost of production" is achieved at the two- to four-man level. As farms get bigger than that, per-unit costs increase.

The reasons for this are both organizational and, surprisingly enough, physical. In the past fifty years, as farm implement design has improved, average tractor size has steadily risen. The same farmer, driving increasingly large-sized equipment, covers more and more area. There seemed until recently to be no limit to increasing human productivity. Each subsequent horsepower cost less than the one before, and in agricultural areas flat and continuous enough to justify it, farmers traded equipment again and again to keep up with the largest units available.

In recent years, though, the limit seems to have been reached. Four-wheel-drive tractors of two to three hundred horsepower, articulated in the middle to cut down turning radius, seem about as large as tractors ought to go, even in the areas with the broadest terrain. The limitation on tractor size is, true to the name of the beast, traction. Farmers have tried still larger tractors, and there is no way to use the power they afford. The wheels slip. Add more wheels, and one runs out of crop room and turning room. Gravity and friction conspire to keep good growing on a human scale.

Angelo Mazzei, the engineer who once ran TAP's maintenance shop, said, "We tried two-hundred-seventy-five-horsepower rigs and some experimental machines about twice as large. The big ones did a little more work, but the wheels slipped so much they wore out tires so fast they didn't pay. They tied up capital, too — when they were down, we were out of a whole lot of work. And they tended to compact the soil even more than the two-seventy-fives."

Once a farm is of a size to use the largest equipment that can be drawn by the largest viable tractor, the next step to increasing farm size is to have two of these sets of largest tractor and largest equipment to go with it. And once a farmer gets much beyond two or three blocs of this nature, he is so busy with organizational chores of managing personnel, allocating work and capital, marketing, hiring, and firing that he is no longer driving one of the machines. It is at this point that field efficiencies begin to decrease.

An economist writing recently in the *Farm Index*, journal of the Economic Research Service of the U.S. Department of Agriculture, suggested some sample optimal sizes of farms. The acreages indicated are considerably larger than the average American farm of today, which has about 380 acres of land. An Indiana grain farm, raising 800 acres of corn and soybeans, a Montana wheat and barley farm of about 2,000 acres, a Louisiana rice and soybean farm of about 360 acres, a Mississippi Delta cotton and soybean farm of about 600 acres all enjoy the lowest unit costs of production. The obvious point to be learned here is that even though these "optimal-sized farms" are quite large, they are far smaller than Tejon Ranch or even than the Tejon Agricultural Partners farm. "When it comes to buying production items," the *Farm Index* states, "and to selling their products, large-scale farmers do enjoy some price advantages. But the evidence suggests that these advantages, where they exist, tend to be minor. In themselves they don't provide sufficient reasons for farm enlargement. . . ."

What is surprising is that even in the world of upper management of giant-sized corporate farms, the true nature of optimal-sized farms seems to be well known. When pressed, a surprising array of managers made comments that might lead one to suspect that forces other than efficient capital utilization on the farm are responsible for the size of TAP.

The well-thought-of farm manager, Karl Fanucchi, feels the twenty-thousand-acre farms he has run are larger than an efficient farm should be.

"I think twelve hundred acres is the most efficient size for cotton/grain out here. Double that size for level land. One man can only watch so many laborers. As soon as a farmer has to hire a foreman, he starts to lose efficiency. TAP and River West farms both are too big in terms of efficiency. We have slippage — for example, before, there were eight laborers standing around while we adjusted those jammed harrows. But the financial institutions won't back the size operation that's cheapest. Don't ask me why, but that sort of farm has a tough time until there's a paid-for home piece to use as collateral. You got to keep growing so the bank doesn't know what you're up to. It's wrong, but it's a fact of life here — financing."

John Wood, current general manager of TAP, grew up on a family-sized sugar cane farm in Louisiana. "In cane," he says, "the most efficient size is probably only about five hundred acres' worth. That's production for one harvester, one loader, et cetera. All the equipment components fit. A single manager can oversee it, not overburdened, but not wasting time. At TAP we're going to end up with about ten thousand acres of trees and vines. I do think ten thousand acres is big for the valley."

Howard Leach, former president of Tejon Ranch Corporation, would not be pressed on the matter, but did comment, "If you were a limited partner, how'd you like to hear that your farm was over the maximum level of efficient production?"

And Tim Heinrichs, the agricultural journalist with years of experience covering the southern Valley, when asked about farm size, said, "The optimal-sized farm here would be fairly large, and it would be noncorporate. I'd say eight hundred or nine hundred acres of vegetables. Anywhere in the six-hundred-to-two-thousand-acre range."

Ray Peterson, the economist, once said, "As you move from one man with a small operation doing the work himself, and doing the accounting himself, you incur the cost of professional managers. Also, a big farm can be too big for everyone to know everything. This sometimes causes operations to go wrong for a while. There was a case recently here at TAP in which the wrong pesticide was applied to eighty or a hundred and sixty acres, and no one stopped the operator. Economies of scale do stop when you get to the largest production bloc. After that, you proceed in multiples of these units. At a certain point, the cost of administering the multiples decreases the advantage of further expansion."

Why then did TAP set out to farm twenty-one thousand acres of land? Ray Peterson commented, "Marketing advantages continue to develop as farm size increases. Processing plants today are highly capitalized and will pay premium prices for an assured source of supply."

"Size is leverage. If you were an almond huller," says John Wood, "you would much rather tie up this acreage and deal with one farm than deal with one or two hundred temperamental guys growing small acreages down the road and wanting all different things and contracts. The savings in the cost of doing business would justify paying premium prices to us. With our size we also save on inputs. We get twelve percent off on tractors, which is something only the giants get. We save more than others on implements and on our herbicides and pesticides; we deal with one company that won't even sell to small individual farmers, because it's not the kind of business where they feel they make money."

Very large farms that are well run do make excellent use of their machinery. According to many studies, their tractors and equipment work more hours each year than those of family farmers. They may purchase the most specialized of timesavers, such as Jimmy O'Brien's post-driver truck. They may hire experts, economists such as Ray Peterson, or Angelo Mazzei, who has an M.A. in mechanical engineering, and, perhaps more to the point, lawyers to help managers discern business strategies that make best use of the tax laws.

Kenneth Krause and Leonard Kyle, writing in the *American Journal of Agricultural Economics*, comment that

... tax loss farms are a spur to large scale growth that needs investigating. . . . No longer is the belief tenable that weather, biological processes, and the superior incentives of unpaid family members provide impossible barriers to a large scale industrialized agriculture . . . the gradual industrialization of production is being forced by industrialization of the processing, handling and distribution of food . . . part of the incentive for expansion and formation of very large farm units comes from outside the traditional farm sector. With the large amounts of capital accumulation in the U.S. in the last ten years, many high-income individuals and growth-oriented business firms are looking for new ventures.

When they risk money in farming ventures investors bet that farm products will continue to

... have a growing demand, require relatively low inputs of unorganized labor and provide opportunities for large scale control and integration of production, processing, marketing, and advertising promotions. . . . Control of large blocks of farmland . . . preempts use by competitors.

The land farmed remains available for future nonagricultural development. Investors, with help from taxpayers, bear the holding costs. If farms grow large enough, if the tax environment and the terms under which credit becomes available and the requirements of food marketing grow so stringent that conditions favor the largest operators, there is, of course, an eventual danger of an oligopolistic food supply situation developing at the production level. But it's not likely soon. There are still too many farmers who capitalized at preinflation prices and who farm with breathtaking efficiency on a relatively small scale.

This is certainly not to say that the control of the nation's food supply exercised by large corporations is decreasing, for that is not true. Such control is already enormous, and it is in fact increasing. The inputs needed by farmers are already under corporate control. Most American tractors, chemicals, implements, breeding stock, work clothes, most truck dump mechanisms, tractor tires, spray nozzles, veterinary supplies, feed grain, and most of the hundreds of other ingredients that go into making a farm run at a modern level of productivity, come from economically concentrated suppliers. Even the beer farmers drink on break comes from transnational giants that have crowded local brewers from the scene.

Until recently, the same thing has not been as true on the output end of the farmers' production. It has been a long time since the noble yeoman farmer, after growing all that the family needed, could hitch up the team and take a wagonload of whatever was left over down to trade at the local ma-and-pa grocery. But there were until recently more opportunities in key farming areas to market farm production than now remain — auctions, competing agents, institutions bidding for direct purchase. More and more food is sold either by chain restaurants for consumption away from home, or by chain markets for consumption at home. The marketplace is growing organized. Central buying by retailers leads to centralized wholesaling by processors. There are only five major grain wholesalers. The more processed foods have already moved into oligopolistic marketing situations. Less than free market conditions are involved in the sale of nearly everything preserved that is stocked in supermarkets. The experience of supermarketing in Massachusetts, Iowa, and California, and finding the same brands from the same suppliers on sale coast to coast brings home the message — the farmer is the last free market link in a food production chain increasingly dominated in input, processing, and marketing aspects by a few powerful financial forces.

Smaller farmers faced with this sort of supply system and marketplace either switch to less perfectly supplied crops, or respond by banding together to form purchasing co-ops and marketing organizations. They organize to protect themselves — rather like unionizing farm workers.

Until now, there have always been too many farmers to permit them effective bargaining, except for a few short-run dramas. Nowadays, although there are still two million two hundred thousand farms in America, the likelihood of organizing crop marketing organizations has increased. Half of U.S. farm production is generated by only 6 percent of farms. For the first time marketing organizations of few enough and disciplined enough members to swing actual power in regional marketplaces exist. Nut growers have organized; cannery tomato growers have organized; citrus growers, cranberry growers, rice growers, avocado growers have all organized. The list is long.

According to a 1975 article in the *Farm Index,*

> Fewer and fewer firms will control vegetable marketing in the 1980s. A major reason is the increased organization in moving fresh vegetables over long distances.
>
> A half century ago, vegetable suppliers were located near major cities in the East. Today, most vegetables come from the West. Although informal agreements between large buyers in the East and large suppliers in the West are hardly new, the Economic Research Service researchers suggest that a trend to formalize such agreements may be coming in the form of integrated membership on boards of directors.
>
> In this form, a member of a board of directors of a large retailer may hold a large number of shares in a food supply firm and sit also on that firm's board of directors. . . .
>
> Another form of control occurs when one big supplier becomes the sole source of supply for a food chain's market needs for one or several vegetable crops.
>
> In a similar development, processing vegetable growers are also becoming more organized. Years ago, a purchaser could approach them individually, playing them off. . . . Now, many grower-processors have formed organizations to bargain collectively. . . .

Along with large processors and large marketing chains and large input suppliers come organizations of growers. It may look

like a just dance. It may appear that each force calls into existence its own "countervailing force," that each of the resulting giants is the nemesis of the giant it buys from and the giant to which it sells. But such is not the case if their managements and financing are integrated. Although they are overtly autonomous, it's revealing to see these large links in the food supply chain not as adversaries but as contending elements in a great "economic bureaucracy" (as economist Richard Goodwin calls it in *The American Condition*) whose internal competition serves only to keep it trim.

Surprisingly, from the point of view of the consumer it may not make much difference if big companies own farms or if smaller farms band together to market goods or join with processors to share the risks of growing and marketing. The sort of food pricing and food quality available in any of these eventualities will leave the consumer the poorer. Pure competition is a tasty and a thrifty spice. Its replacement by the vast supply structure that grows ever larger not only drives small manufacturers, small suppliers, small farmers, small processors, small shippers and small merchants out of jobs holding promise of self-satisfaction, it also changes the nature of our political environment drastically.

In impoverished countries, the power of food is manifest in venomous ways — allegiance is owed to a boss, a ruler or creed that offers food for the table. In America, nothing so vulgar happens regularly.

Here, until recently, the broad base of supply has kept the consumer in charge of the marketplace. The danger of oligopolistic control of the food supply is obvious — once able to charge enough to assure themselves of excessive profit, oligopolists amass capital that is used to further increase control and stifle competition.

It is also true that an artificially high price level can accrue to companies that are very large but have less than an oligopolistic share of the marketplace — companies large enough to advertise nationally, large enough to dominate regional or seasonal supply

situations, large enough to have access to privileged contracts or financing resources.

The result of our food supply's shifting into concentrated control is that food costs more, and that the profits from selling it essentially go to finance further shifts in control. For the consumer, this is now the stuff of annoyance. It provides, during shopping trips through the supermarket, an unjoyful background of nagging indignation. The shifts of price are tiny — an unjust penny or two here and there, for some processed foods. But the shifts in power that these shifts in price represent and help finance are serious threats to our collective independence. In the only slightly distant future, when it becomes more obvious to more people that something precious has been removed, that there's nothing useful left for most of us to do, and when, as remedy, we wish to start saying "no" to some technology and some business structures, then let us hope that we can bring the political power into our hands to do so.